Performance Analysis and Improvement in MIMO Communication Systems

Xuhui Ding · Dekang Liu · Ziyi Yang ·
Gaoyang Li · Neng Ye · Kai Yang
Editors

Performance Analysis and Improvement in MIMO Communication Systems

Springer

Editors
Xuhui Ding
School of Cyberspace Science and Technology
Beijing Institute of Technology
Beijing, China

Ziyi Yang
School of Cyberspace Science and Technology
Beijing Institute of Technology
Beijing, China

Neng Ye
School of Cyberspace Science and Technology
Beijing Institute of Technology
Beijing, China

Dekang Liu
Institute of Unmanned System
Beihang University
Beijing, China

Gaoyang Li
School of Biological Science and Medical Engineering
Beihang University
Beijing, China

Kai Yang
School of Information and Electronics
Beijing Institute of Technology
Beijing, China

ISBN 978-981-96-5950-0 ISBN 978-981-96-5951-7 (eBook)
https://doi.org/10.1007/978-981-96-5951-7

This Springer imprint is published by the registered company Springer Nature Singapore Pte Ltd.
The registered company address is: 152 Beach Road, #21-01/04 Gateway East, Singapore 189721, Singapore

Preface

The book focuses on modeling and analyzing large-scale array communication systems to solve the computational complexity problems caused by high-dimensional arrays. This is achieved by providing an in-depth study on several major topics, such as channel estimation, delay estimation, angle estimation, and joint angle-delay estimation. Various technologies that improve the performance of networks and achieve multiple-input multiple-output (MIMO) communications are discussed in this book. In Chap. 1, we introduce the basic knowledge of MIMO and the book's structure. In Chap. 2, we have investigated the training optimizations for hybrid MIMO communications. Channel estimation is conceived for hybrid MIMO communication systems. Both mean square error minimization and MI maximization are used as our performance metrics, and a pair of low-complexity channel estimation schemes are proposed. In Chap. 3, we have investigated the problems of accurate time delay estimation, the design of training pilots, and hybrid matrix optimization within the large-scale antenna array terahertz (THz) broadband communication system. In Chap. 4, we have investigated the problem of the angle of arriving tracking for the large-scale array in THz communication, which has a large size and a narrow beam, highly demanding an accurate angle estimation. In Chap. 5, we investigate a joint angle-delay estimation with a large-scale array in the THz communication system. In Chap. 6, we have investigated the two-timescale robust system energy efficiency (EE) precoding design for the multi-user dual-polarized MIMO system in depth. In Chap. 7, we have investigated the EE optimization problem of MIMO multi-satellite uplink communication systems based on lens antenna arrays. In the final chapter, we conclude the book and propose future directions.

Beijing, China
December 2024

Xuhui Ding
Dekang Liu
Ziyi Yang
Neng Ye
Gaoyang Li
Kai Yang

Contents

About the Editors

Xuhui Ding received his B.E. and Ph.D. degrees from Beijing Institute of Technology, China, in 2014 and 2021, respectively. From July 2019 to September 2019, he was a visiting student with the Chalmers University of Technology, Göteborg, Sweden. He is currently an Assistant Professor with the School of Cyberspace Science and Technology, Beijing Institute of Technology (BIT). His research interests include MIMO communication, iterative receiver design, error control coding and information theory, spectrum sensing for cognitive radio, and deep learning. He is a lead Guest Editor of Electronics from 2022 to 2024. He has been an Editorial Board Member of Transactions of NUAA and Journal of Data Acquisition and Processing.

Dekang Liu received his B.E. and Ph.D. degrees from Beijing Institute of Technology, China, in 2013 and 2020, respectively. From November 2020 to March 2024, he was a postdoctor with the Intelligent Collaborative Perception and Analysis Cognitive Laboratory, Beijing. He is currently an Associate researcher with the Institute of Unmanned System, Beihang University. His research interests include wireless communication, channel estimation, integrated sensing and communications.

Ziyi Yang received the Ph.D. degree in electronic engineering from Beijing Institute of Technology, Beijing, China, in 2021. She is currently with the School of Cyberspace Science and Technology, Beijing Institute of Technology, China, as a special associate researcher. Her research interests include wireless communication, aerospace communication, green communication, signal processing, and convex optimization.

Gaoyang Li received his Ph.D. degree from Tohoku University, Japan in 2020. He then did postdoctoral research at the Japan Society for the Promotion of Science (JSPS), Institute of Fluid Science, Tohoku University. His research interests include artificial intelligence, deep learning, data science, and biomedical engineering.

Neng Ye received the B.S. and Ph.D. degrees from the Beijing Institute of Technology, China, in 2015 and 2021, respectively, where he is currently an Assistant Professor with the School of Cyberspace Science and Technology. He has contributed more than 40 peer-reviewed articles including three ESI highly cited papers and one best paper reward of IWCMC 2021. He has also authored two monographs published by Springer Nature and been granted more than 10 invention patents, including five US patents. He received the Best Ph.D. Thesis Award of China Education Society of Electronics in 2022 and the Young Elite Scientists Sponsorship Program by China Association for Science and Technology in 2022. He has been an Editor of the Journal of Electronics and Information Technology since 2021. He was a Lead Guest Editor of Sensors and Electronics from 2021 to 2023. His research interests include non-terrestrial wireless communications, multi-user communications, and intelligent signal processing.

Kai Yang received the B.E. degree in communications engineering from the National University of Defense Technology, Changsha, China, in 2005 and the Ph.D. degree in communications engineering from the Beijing Institute of Technology, Beijing, China, in 2010. From January 2010 to July 2010, he was with the Department of Electronic and Information Engineering, Hong Kong Polytechnic University, Hong Kong. From 2010 to 2013, he was with Alcatel-Lucent Shanghai Bell, Shanghai, China. In 2013, he joined the Laboratoire de Recherche en Informatique, University Paris Sud 11, Orsay, France. He is currently a Full Professor with the School of Information and Electronics, Beijing Institute of Technology. His current research interests include convex optimization, massive MIMO, mmWave systems, resource allocation, and interference mitigation.

Acronyms

5G	Fifth-Generation
6G	Sixth-Generation
ADC	Analog-to-Digital Converter
AI	Artificial Intelligence
AoA	Angle of Arrival
AWGN	Additive White Gaussian Noise
BPSK	Binary Phase Shift Keying
BS	Base Station
CIRs	Channel Impulse Responses
CRLB	Cramer-Rao Lower Bound
CS	Compressed Sensing
CSI	Channel State Information
CSIT	Channel State Information at the Transmitter
DAC	Digital-to-Analog Converter
DFT	Discrete Fourier Transform
DL	Deep Learning
DoA	Direction of Arrival
EE	Energy Efficiency
EKF	Extended Kalman Filter
EM	Expectation Maximization
EVD	Eigen Value Decomposition
FFT	Fast Fourier Transform
GAMP	Generalized Approximated Message Passing
GEO	Geostationary Orbit
GHz	Gigahertz
i.i.d.	Independent and Identically Distributed
KKT	Karush-Kuhn-Tucker
LEO	Low Earth Orbit
LMI	Linear Matrix Inequality
LMMSE	Linear Minimum Mean Square Error
LoS	Line-of-Sight

LS	Least Squares
MI	Mutual Information
MIMO	Multiple-Input Multiple-Output
ML	Maximum Likelihood
MLE	Maximum Likelihood Estimation
MMSE	Minimum Mean Square Error
MMSE-DFT	Minimum Mean Square Error Estimator with Orthogonal Probing Based on Discrete Fourier Transform
MMSE-ITE	Iterative Minimum Mean Square Error
mmWave	Millimeter Wave
MSE	Mean Square Error
MU	Mobile User
MUSIC	Multiple Signal Classification
NLoS	Non-Line-of-Sight
NOMA	Non-Orthogonal Multiple Access
OFDM	Orthogonal Frequency Division Multiplexing
OMP	Orthogonal Matching Pursuit
OOK	On-Off Keying
PSK	Phase Shift Keying
RF	Radio Frequency
RS	Rate-Splitting
SAA	Simulated Annealing Algorithm
SDP	Semidefinite Programming
SE	Spectrum Efficiency
SINR	Signal to Interference Plus Noise Ratio
SLS	Scaled Least Squares
SNR	Signal-to-Noise Ratio
SVD	Singular Value Decomposition
TDL	Time Delay Line
THz	Terahertz
UAV	Unmanned Aerial Vehicle
WMC	Wideband Modulation Converter

Chapter 1
Introduction

Kai Yang, Dekang Liu, and Xuhui Ding

This chapter first introduces the significance of performance analysis and improvement in multiple-input multiple-output (MIMO) communication systems. Then, two aspects of performance analysis and improvement, including parameter estimation and energy efficiency (EE) optimization are discussed. Finally, the organization of this book is presented.

1.1 Developments of MIMO Communications

With the advent of the fifth-generation (5G) mobile communication era, MIMO systems, as a key technology in 5G [1, 2], has attracted wide attention [3–7]. Research has shown that MIMO systems can achieve extremely high beamforming gains [4], effectively enhancing received signal power [8]. Moreover, it can leverage the advantages of narrow beams to simultaneously serve multiple users, thus improving spectrum efficiency (SE) [8, 9].

K. Yang · X. Ding (✉)
Beijing Institute of Technology, Beijing, China
e-mail: dingxuhui@bit.edu.cn

K. Yang
e-mail: yangkai@bit.edu.cn

D. Liu
Beijing Institute of Technology, Beijing, China
e-mail: liudekang@buaa.edu.cn

X. Ding et al. (eds.), *Performance Analysis and Improvement in MIMO Communication Systems*, https://doi.org/10.1007/978-981-96-5951-7_1

To meet the growing demands of communication and achieve high-performance MIMO communication systems, various promising technologies should be investigated. Specifically, the parameter estimation problem in MIMO communication systems needs to be addressed. Large-scale arrays often imply high-dimensional parameter estimation, leading to extremely high computational complexity. Moreover, the performance of MIMO systems is closely related to the accuracy of parameter estimation, including parameters such as channel [10, 11], delay [12], and angle [13]. Therefore, in order to fully exploit the performance advantages of MIMO systems, there is an urgent need to research parameter estimation techniques with high performance and low complexity. Additionally, with the rapid development of wireless communication technology, the energy consumption of communication is increasing. Thus, EE has become a key performance indicator for current and future mobile communication systems [14], improving EE can not only alleviate environmental pressures but also reduce the economic costs of communication operators.

Motivated by the above observations, in this book, we will investigate parameter estimation and EE optimization algorithms for MIMO communication systems, which will provide useful guidance for the design of MIMO systems.

1.2 Basic Concept and Features

1.2.1 Parameter Estimation in MIMO Systems

Parameter estimation is crucial for the implementation of MIMO communication systems.

Firstly, obtaining strict and accurate channel state information (CSI) is a prerequisite for leveraging the enormous performance advantages of MIMO technology [15]. Traditional signal processing in MIMO systems is mainly implemented at baseband, which means that each antenna requires a separate radio frequency (RF) link, leading to complex system structures and high costs. Therefore, low-complexity channel estimation for MIMO systems is a hot research topic. Meanwhile, based on large-scale arrays, using phase shifter networks to combine signals from numerous antennas into a limited number of RF links forms large-scale hybrid arrays. In hybrid structure channel estimation, constrained by the dimensions of the hybrid matrix, it is not feasible to obtain CSI through traditional all-digital methods. To avoid the high computational complexity of high-dimensional matrices, many studies have focused on spatial sparsity, using compressive sensing theory to recover sparse channels [16]. However, for communication scenarios with relatively close communication distances and complex scattering environments, the sparse characteristics of the channel may not be satisfied, and algorithms based on sparse assumptions may be limited.

Secondly, delay estimation is also crucial for MIMO communication systems. Due to the large signal bandwidth and short symbol duration in MIMO systems, there is a high demand for the accuracy of delay estimation [12]. Additionally, distance-selective fading can introduce severe distortion, resulting in signal attenuation and delay spread [17], which poses certain difficulties for accurate delay estimation.

Furthermore, although digital synchronization techniques offer efficient and flexible advantages [18, 19], full digital MIMO systems are constrained by cost and computational complexity, making widespread deployment challenging in the short term. Meanwhile, traditional algorithms based on methods such as cross-spectrum estimation [20], higher-order statistical estimation methods [21], adaptive delay estimation methods [22], and parameter modeling methods [23] in MIMO communication systems face issues such as high computational complexity and limited estimation accuracy, imposing significant pressure on real-time signal processing.

Additionally, angle of arrival (AoA) estimation is a crucial area in MIMO signal processing. The beams in MIMO systems are extremely narrow, making them highly sensitive to angle estimation accuracy [24, 25], thus placing higher demands on the performance of AoA estimation algorithms. Traditional AoA estimation methods mostly target uniform linear array models and utilize all-digital structures. However, in MIMO systems, hybrid structures and sparse arrays are commonly used. The limited number of observations poses certain difficulties for accurate angle estimation, while high-dimensional channels also lead to increased computational complexity. Furthermore, algorithms mostly rely on compressive sensing techniques, which depend on the discretization of spatial angles, thereby limiting estimation accuracy due to grid partition precision. Common angle scanning methods and angle estimation methods based on sparse arrays fail to fully exploit the estimation accuracy advantages brought by high signal-to-noise ratio (SNR).

Therefore, it is necessary to research low-complexity, high-performance parameter estimation algorithms to address the challenges faced by MIMO systems mentioned above.

1.2.2 Energy Efficiency Optimization for MIMO Systems

With the rapid development of wireless communication technology, the energy consumption of wireless communication systems is increasing [26]. Particularly, for MIMO communication systems, the use of multiple antennas also increases the system's power consumption and complexity to a certain extent. Improving EE can not only alleviate environmental pressures [27] but also reduce the economic costs of communication operators. Therefore, EE has become a key performance indicator for current and future mobile communication systems.

With the advancement of antenna technology, dual-polarized antenna arrays [28] and lens antenna arrays have been proposed [29]. In dual-polarized antenna arrays, antennas are paired and grouped according to horizontal and vertical polarizations, and a pair of orthogonally polarized antennas can be configured at the same position, greatly reducing the communication system's demand for equipment spatial dimensions. Additionally, because the electromagnetic waves with orthogonal polarizations experience independent fading during propagation in space, using dual-polarized antenna arrays can achieve lower spatial correlation, effectively reducing the impact of correlation and coupling between antenna elements and lowering the feedback overhead of CSI [30]. Lens antenna arrays consist of an electromagnetic lens with energy convergence properties and a series of matched antennas, and they can replace phase shifter networks to achieve RF analog precoding. They transform the spatial channel into beam space channels associated with directions, where signals from different directions are converged onto different antennas. Due to the sparsity of propagation paths in millimeter wave (mmWave) channels, the beam space channel is sparse, allowing us to select a small number of beams to reduce the dimensionality of MIMO communication systems and the number of RF links [31]. Therefore, these two types of antennas can effectively alleviate the high power consumption problem in future communication systems.

Signal processing, as the primary implementation method for energy optimization, includes resource allocation [32], precoding [33] and antenna selection [25], which need to be designed according to the actual scenarios [34]. However, most research on the above-mentioned antenna array structures has focused on SE. Only comparisons of EE performance have been made between signal processing algorithms based on rate maximization and traditional structures. Therefore, algorithm design directly targeting EE maximization for these two types of antenna arrays holds significant research value.

1.3 Organization

To analyze and optimize the MIMO communication system, this book discusses the parameter estimation algorithms in Chaps. 2–5, and EE optimization methods in Chaps. 6–7. The rest of this book is specified as follows.

Chapter 2 discusses low-complexity estimation algorithms for MIMO communication systems. MIMO communication system features high-dimensional channels, leading to complexity in estimation algorithms. Therefore, we abandon the traditional assumption of channel sparsity in research and instead investigate a universally applicable low-complexity estimation algorithm based on the general MIMO channel model. To design the optimal estimation algorithm, this chapter considers two performance criteria: the minimum mean square error and the maximum mutual information (MI) criterion. To reduce computational complexity, we present two low-complexity channel estimation schemes: one targeting the high-dimensional original channel matrix and the other addressing lower-dimensional composite channel matrices.

Chapter 3 discusses delay estimation techniques in MIMO systems. When the array length of the MIMO receiver is sufficiently long and a broadband signal impinges on the array, the maximum delay spread of the signal reaching the array may greatly exceed the symbol period, resulting in significantly reduced signal correlation across the antennas at the same moment. Although the optimized design of the phase shifter matrix can make the array channel better suited for signal transmission, it does not alter the premise of weak signal correlation entering the combiner. Thus, the array's ability to enhance signal merging is greatly compromised. Therefore, we investigate the delay estimation problem in MIMO systems. Starting from the issue of array signal delay estimation, this chapter proposes a hybrid structure based on the time delay line (TDL) and studies the performance of MIMO receivers based on the TDL hybrid structure.

Chapter 4 discusses angle estimation techniques in MIMO systems. High-gain sharp beams demand greater accuracy in angle estimation, as even slight deviations can severely degrade communication quality. As a result, angle estimation techniques have always been a focal point in research related to MIMO systems. Meanwhile, although hybrid structure MIMO systems reduce costs and computational complexity, the loss of information during dimensionality reduction increases the difficulty of accurately estimating the arrival angles. Therefore, we investigate the feasibility and methods of high-precision angle estimation in hybrid structure MIMO systems by investigating the mechanism of waveform distortion in the time-frequency domain of synthesized signals.

Chapter 5 discusses the joint angle delay estimation in MIMO systems. MIMO systems demand high precision in angle and delay estimation, yet they face challenges of high signal dimensionality and computational complexity. Moreover, MIMO systems exhibit widening effects for high-speed signals, rendering traditional array models inadequate to meet the requirements of high-precision angle estimation. Additionally, broadband signals in high-frequency bands require extremely high sampling rates, imposing significant pressure on real-time signal processing for precise delay estimation. Therefore, we investigate the high-precision, low-complexity joint angle delay estimation in MIMO communication systems, proposing maximum likelihood estimation (MLE) methods based on grid partitioning and off-grid asymptotic Cramer-Rao lower bound (CRLB) estimation methods.

Chapter 6 discusses the energy-efficient robust precoding design method for MIMO communication systems. MIMO communication system based on a dual-polarized antenna array can reduce the influence caused by correlation and coupling between antenna array elements and can reduce the feedback cost of CSI in communication. However, most of the research work takes spectral efficiency as the performance index and lacks the research on EE as the performance index. Therefore, in order to optimize the EE of a multi-user MIMO downlink communication system with double-polarized antenna, we investigate robust precoding design with high energy efficiency.

Chapter 7 discusses the EE optimization in MIMO communication systems based on lens antenna arrays. A lens antenna array comprises an electromagnetic lens with energy-converging capabilities and a series of matched antennas. It can replace phase shifter networks to achieve RF analog precoding, thereby reducing system power consumption. Therefore, we integrate the characteristics of lens antenna array responses and propose beam selection, digital baseband precoding design methods, and high-efficiency RF link number design methods with the objective of maximizing EE.

Chapter 8 summarizes this book and discusses the future direction to analyze and optimize MIMO communication system.

References

1. Andrews JG, Buzzi S, Choi W et al (2014) What will 5G be? IEEE J Sel Areas Commun 32(6):1065–1082
2. Chin WH, Fan Z, Haines R (2014) Emerging technologies and research challenges for 5G wireless networks. IEEE Trans Wirel Commun 21(2):106–112
3. Hoydis J, ten Brink S, Debbah M (2013) Massive MIMO in the UL/DL of cellular networks: How many antennas do we need? IEEE J Sel Areas Commun 31(2):160–171
4. Larsson EG, Edfors O, Tufvesson F et al (2014) Massive MIMO for next generation wireless systems. IEEE Commun Mag 52(2):186–195
5. Jin S, Wang X, Li Z et al (2014) Zero-forcing beamforming in massive MIMO systems with time-shifted pilots. In: IEEE ICC. IEEE, pp 4801–4806
6. Dai Y, Dong X (2016) Power allocation for multi-pair massive MIMO two-way AF relaying with linear processing. IEEE Trans Wirel Commun 15(9):5932–5946
7. Marzetta TL (2010) Noncooperative cellular wireless with unlimited numbers of base station antennas. IEEE Trans Wirel Commun 9(11):3590–3600
8. Heath RW, González-Prelcic N, Rangan S et al (2016) An overview of signal processing techniques for millimeter wave MIMO systems. IEEE J Sel Areas Commun 10(4):436–453
9. Swindlehurst AL, Ayanoglu E, Heydari P et al (2014) Millimeter-wave massive MIMO: the next wireless revolution? IEEE Commun Mag 52(9):56–62
10. Soysal A, Ulukus S (2010) Joint channel estimation and resource allocation for MIMO systems-part I: single-user analysis. IEEE Trans Wirel Commun 9(2):624–631
11. Soysal A, Ulukus S (2010) Joint channel estimation and resource allocation for MIMO systems-part II: multi-user and numerical analysis. IEEE Trans Wirel Commun 9(2):632–640
12. Candes EJ, Fernandez-Granda C (2012) Super-resolution from noisy data. J Fourier Anal Appl 19(6):1229–1254
13. Ge Y, Zhang W, Gao F et al (2019) Angle-domain approach for parameter estimation in high-mobility OFDM with fully/partly calibrated massive ULA. IEEE Trans Wirel Commun 18(1):591–607
14. Wu, J, Rangan S, Zhang H (2016) Green communications: theoretical fundamentals, algorithms, and applications. CRC Press
15. Zhang P, Chen S, Hanzo L (2015) Two-tier channel estimation aided near-capacity MIMO transceivers relying on norm-based joint transmit and receive antenna selection. IEEE Trans Wirel Commun 14(1):122–137
16. Xie H, Gao F, Zhang S et al (2016) UL/DL channel estimation for TDD/FDD massive MIMO systems using DFT and angle reciprocity. In: IEEE 83rd vehicular technology conference (VTC Spring), pp 1–5. IEEE
17. Han C, Bicen AO, Akyildiz IF (2015) Multi-ray channel modeling and wideband characterization for wireless communications in the terahertz band. IEEE Trans Wirel Commun 14(5):2402–2412
18. Joon-Yong Lee RA (2002) Scholtz, Ranging in a dense multipath environment using an UWB radio link. IEEE J Sel Areas Commun 20(9):1677–1683
19. Aedudodla SR, Vijayakumaran S, Wong TF (2005) Timing acquisition in ultra-wideband communication systems. IEEE Trans Veh Technol 54(5):1570–1583
20. Maher PS, Malaney RA (2009) A novel fingerprint location method using ray-tracing. In: GLOBECOM IEEE global telecommunication conference. IEEE, pp 1–5
21. Lin C, Fang W (2013) Joint angle and delay estimation in frequency hopping systems. IEEE Trans Aerosp Electron Syst 49(2):1042–1056
22. Makki A, Siddig A, Saad M et al (2016) Indoor localization using 802.11 time differences of arrival. IEEE Trans Instrum Meas 65(3):614–623
23. Wu S, Zhang Q, Fan R et al (2018) Match-filtering based ToA estimation for IR-UWB ranging systems. Wirel Commun Mob Comput Conf 1099–1105
24. Pal P, Vaidyanathan PP (2010) Nested arrays: a novel approach to array processing with enhanced degrees of freedom. IEEE Trans Signal Process 58(8):4167–4181

25. Ibrahim M, Ramireddy V, Lavrenko A et al (2017) Design and analysis of compressive antenna arrays for direction of arrival estimation. Signal Process 138(9):35–47
26. Zhang S, Cai X, Zhou W et al (2019) Green 5G enabling technologies: an overview. IET Commun 13(2):135–143
27. Auer G, Giannini V, Godor I et al (2011) How much energy is needed to run a wireless network? IEEE Wirel Commun 18(5):40–49
28. Gong S, Xing C, Chen S et al (2017) Secure communications for dualpolarized MIMO systems. IEEE Trans Signal Process 65(16):4117–4192
29. Gao Y, Khaliel M, Zheng F et al (2017) Rotman lens based hybrid analog digital beamforming in massive MIMO systems: array architectures, beam selection algorithms and experiments. IEEE Trans Veh Technol 66(10):9134–9148
30. Dong L, Choo H, Heath RW et al (2005) Simulation of MIMO channel capacity with antenna polarization diversity. IEEE Trans Wirel Commun 4(4):1869–1873
31. Zeng Y, Zhang R, Chen ZN (2017) Electromagnetic lens focusing antenna enabled massive MIMO: performance improvement and cost reduction. IEEE J Sel Areas Commun 32(6):1194–1206
32. Zappone A, Sanguinetti L, Bacci G et al (2016) Energy efficient power control: a look at 5G wireless technologies. IEEE Trans Signal Process 64(1):1668–1683
33. Belmega EV, Lasaulce S (2010) Energy efficient precoding for multiple antenna terminals. IEEE Trans Signal Process 59(1):329–340
34. Zappone A, Jorswieck E (2015) Energy efficiency in wireless networks via fractional programming theory. Found Trends Commun Inf Theory

Chapter 2
Channel Estimation for Hybrid MIMO Communication Systems

Jiawen Chen, Dekang Liu, and Xuhui Ding

In this chapter, we propose a pair of low-complexity channel estimation schemes. Section 2.1 introduces the research status of estimating large-scale MIMO channels. Section 2.2 gives the signal model for channel estimation in hybrid MIMO systems. In Sect. 2.3, left-right channel estimation schemes are investigated. Then, a composite channel estimation is presented in Sect. 2.4. The simulation results are given in Sect. 2.5 to assess the performance of the proposed channel estimation algorithms. Our conclusions are drawn in Sect. 2.6.

2.1 Introduction

Multiple antenna-aided MIMO techniques [1–4] are recognized as one of the important pillars of next-generation wireless networks. In the quest for increased degrees of freedom, the scale of MIMO systems has been increasing as they have evolved. Massive MIMO or large-scale MIMO schemes constitute an important enabling technology, which has already been widely used in certain communication applications. For example, on satellites, the number of antennas may be as high as 300. Furthermore, ships and other large-bodied vehicles have also used large-scale antenna arrays. However, the potential performance gains are critically dependent on the availability

J. Chen · X. Ding (✉)
Beijing Institute of Technology, Beijing, China
e-mail: dingxuhui@bit.edu.cn

J. Chen
e-mail: chenjiawen@bit.edu.cn

D. Liu
Beihang University, Beijing, China
e-mail: liudekang@buaa.edu.cn

X. Ding et al. (eds.), *Performance Analysis and Improvement in MIMO Communication Systems*, https://doi.org/10.1007/978-981-96-5951-7_2

and accuracy or the absence of CSI. Therefore, channel estimation becomes a critical part of various MIMO communication systems [5–12].

Generally, a classic channel estimation task consists of two components, training optimization and channel estimator design. There is a rich body of literature on channel estimation. When the channel statistics are unknown, the least squares (LS) channel estimator is preferred and random training sequences having a white spectrum constitute the optimal choice. By contrast, when the channel statistics are known, the maximum likelihood (ML) or the linear minimum mean square error (LMMSE) channel estimator [12–14] are the natural choices. However, in this case, the training optimization is more complicated. Moreover, when specific constraints have to be taken into account, such as the per-antenna power constraints, the training optimization becomes even more complex [15].

The channel estimation problem of massive MIMO systems has been extensively studied in the literature [11, 16–20]. The key behind these contributions is how to exploit the specific structure of massive MIMO channels for reducing the escalating overhead for channel estimation [11, 20–25]. For mmWave-based massive MIMO systems, the sparsity of the mmWave channel impulse responses (CIRs) has been exploited in [11, 22]. By reconsidering channel estimation with special emphasis on the angular domain [21–23], the resultant channel estimation algorithms become capable of improving the estimation accuracy, despite their reduced overhead.

For large-scale MIMO systems, the popular family of hybrid structures constitutes an economic way of striking a performance vs. complexity trade-off [26, 27]. Explicitly, a hybrid MIMO structure is composed of an analog part and a digital part [26]. In the analog part, only signal phases are adjusted by analog phase shifters. Although these hybrid structures strike a compelling performance vs. complexity trade-off for large-scale MIMOs, they impose new challenges. In particular, because of the nonconvex nature of optimizing the analog part [26], the holistic optimization of hybrid MIMO systems is more challenging than that of their fully digital counterparts [27]. Since the knowledge of the CIRs is required, channel estimation is an essential prerequisite for hybrid transceiver designs [27]. It is worth noting that channel estimation must be performed in the digital domain. Explicitly, channel estimation conceived for hybrid MIMO channels is much more challenging than for its fully digital counterpart.

Moreover, many hybrid MIMO communication systems operate at microwave frequencies [28, 29], which are lower than mmWave carriers. The challenge is that their CIRs are typically not sparse. In this case, we have to estimate large-scale MIMO channels when conceiving hybrid MIMO systems. This is more challenging than traditional channel estimation owing to the constant modulus constraints imposed on the associated analog matrices. Hence the conventional designs are unsuitable for striking an attractive performance vs. complexity trade-off for large-scale hybrid MIMO systems.

Notation: The matrix $\boldsymbol{Z}^{\frac{1}{2}}$ is the Hermitian square root of positive semidefinite $\boldsymbol{Z}$. The identity matrix of appropriate dimension is denoted by $\boldsymbol{I}$, and $\otimes$ denotes Kronecker product. The expressions $\boldsymbol{\Lambda} \searrow$ and $\boldsymbol{\Lambda} \nearrow$ represent a rectangular or square diagonal

Table 2.1 The list of math notations

Variable	Definition
$\boldsymbol{\Sigma}_{\mathrm{H}}, \boldsymbol{\Psi}_{\mathrm{H}}$	The receive and transmit correlation matrices of the MIMO channel $\boldsymbol{H}$
$\boldsymbol{\Sigma}_{\mathrm{N}}, \boldsymbol{\Psi}_{\mathrm{N}}$	The receive and time-domain correlation matrices of the additive noise $\boldsymbol{N}$
$\boldsymbol{X}$	The training sequence to be optimized at the transmitter
$\boldsymbol{G}_{\mathrm{A}}, \boldsymbol{F}_{\mathrm{A}}$	The receive and transmit analog matrices
$\boldsymbol{G}_{\mathrm{D,L}}, \boldsymbol{G}_{\mathrm{D,R}}$	The left and right channel estimators in Sect. 2.3
$\boldsymbol{U}_{\mathrm{DFT}}$	The discrete Fourier transform (DFT) matrix
α_1, β_1	The auxiliary variables in the channel estimation mean square error (MSE) matrix $\boldsymbol{\Phi}_{\mathrm{MSE}}(\boldsymbol{F}_{\mathrm{A}}, \boldsymbol{X})$
P	The maximum transmit power
$\boldsymbol{G}_{\mathrm{D}}$	The minimum mean square error estimator (MMSE) channel estimator in Sect. 2.4
α_2	The auxiliary variable in the composite channel estimation MSE matrix $\boldsymbol{\Phi}_{\mathrm{MSE}}(\boldsymbol{F}_{\mathrm{A}}, \boldsymbol{X}, \boldsymbol{G}_{\mathrm{A}}, \boldsymbol{G}_{\mathrm{D}})$
$\mathrm{vec}(\boldsymbol{Y})$	The vectorization operation stacks the columns of $\boldsymbol{Y}$ to construct a bigger column vector
$\mathrm{Tr}(\boldsymbol{Y})$	The trace of $\boldsymbol{Y}$
$\mathbb{E}\{\cdot\}$	The mathematical expectation

matrix with the diagonal elements in descending order and ascending order, respectively. In addition, $\|\cdot\|_F$ is the matrix Frobenius norm. $[\boldsymbol{Y}]_{:,N}$ consists of the first N columns of $\boldsymbol{Y}$, $[\boldsymbol{Y}]_{[1:N_1;1:N_2]}$ is the sub-matrix consisting of the first N_1 rows and the first N_2 columns of $\boldsymbol{Y}$. For further clarification, some important math definitions in the following derivations are summarized in Table 2.1.

2.2 System Structure and Signal Model

In the system, the transmitter and the receiver are both equipped with hybrid structured antenna arrays. At both the transmitter and the receiver, the number of RF chains is smaller than that of antennas. Due to the hybrid structure, it is impossible to estimate the channel matrix using a constant analog matrix at the transmitter or receiver. To clarify this fact, a brief discussion is first given.

Consider that a training sequence $\boldsymbol{X}_1$ is multiplied by an analog matrix $\boldsymbol{F}_{\mathrm{A},1}$ before being transmitted to the destination. At the destination, the received signal $\boldsymbol{Y}_{\mathrm{F},1}$ is given by

$$\boldsymbol{Y}_{\mathrm{F},1} = \boldsymbol{H}\boldsymbol{F}_{\mathrm{A},1}\boldsymbol{X}_1 + \boldsymbol{N}_1, \tag{2.1}$$

where $\boldsymbol{H}$ is the channel matrix to be estimated and $\boldsymbol{N}_1$ is the additive noise matrix at the destination. The channel matrix satisfies the following Kronecker product structure [30]

$$\boldsymbol{H} = \boldsymbol{\Sigma}_{\mathrm{H}}^{\frac{1}{2}} \boldsymbol{H}_{\mathrm{W}} \boldsymbol{\Psi}_{\mathrm{H}}^{\frac{1}{2}}, \tag{2.2}$$

where each element of $\boldsymbol{H}_{\mathrm{W}}$ is an independent and identically distributed (i.i.d.) Gaussian variable with zero mean and unit variate. The matrices $\boldsymbol{\Sigma}_{\mathrm{H}}$ and $\boldsymbol{\Psi}_{\mathrm{H}}$ are the receive and transmit correlation matrices of the MIMO channel, respectively.
Since channel estimation is performed in the digital domain, the received training signal is first multiplied by an analog matrix $\boldsymbol{G}_{\mathrm{A},1}$ before being processed by the fully digital channel estimator. Therefore, for channel estimation, the received signal in the digital domain is given by

$$\boldsymbol{Y}_{\mathrm{D},1} = \boldsymbol{G}_{\mathrm{A},1} \boldsymbol{H} \boldsymbol{F}_{\mathrm{A},1} \boldsymbol{X}_1 + \boldsymbol{G}_{\mathrm{A},1} \boldsymbol{N}_1. \tag{2.3}$$

Naturally, channel estimation aims to recover the desired channel matrix $\boldsymbol{H}$ from the received signal $\boldsymbol{Y}_{\mathrm{D},1}$. Unfortunately, as illustrated by Scheme 1 in Fig. 2.1, this task is impossible based on (2.3). This is because for hybrid MIMO systems, $\boldsymbol{G}_{\mathrm{A},1}$ is a fat matrix and $\boldsymbol{F}_{\mathrm{A},1}$ is a tall matrix. Thus based on the previous signal model, no matter how long the column of $\boldsymbol{X}_1$, the matrix $\boldsymbol{F}_{\mathrm{A},1} \boldsymbol{X}_1$ is usually a matrix that is rank deficient, and thus it is impossible to estimate the channel matrix accurately. It is worth noting that in some special cases, there may exist some special structures that can be exploited so that channel estimation based on (2.3) is achievable. For

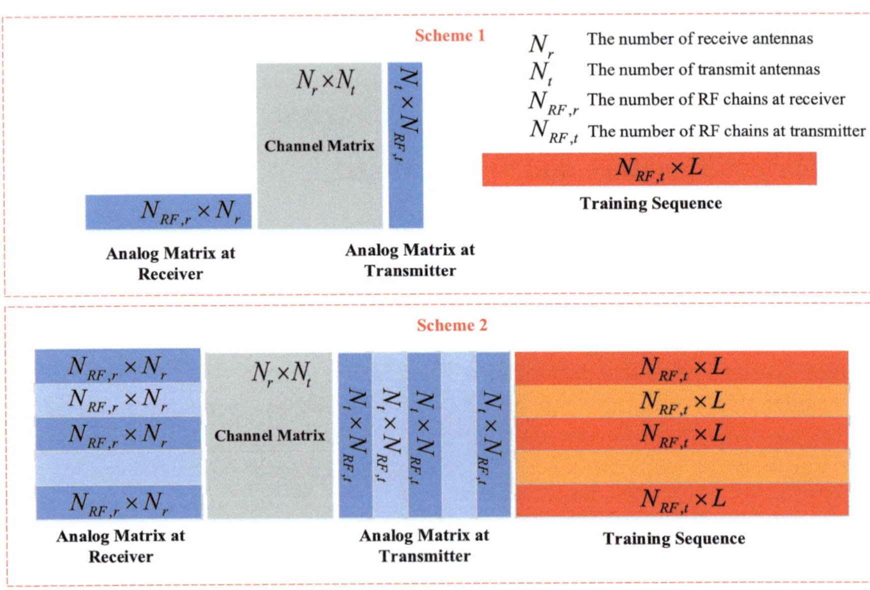

Fig. 2.1 Two potential approaches for channel estimation for hybrid MIMO systems

example, for mmWave channels, the channel matrices exhibit sparsity structures, which means that the number of parameters to estimate, i.e., the true dimension of the MIMO channel matrix, is much smaller than the size of the channel matrix. Then with the aid of compressed sensing (CS) techniques, channel estimation is achievable. However, this result does not hold for the general channel matrix, e.g., micro-wave channels. In this work, we focus on channel estimation for a general large dimensional channel matrix, where sparsity property does not exist.

Thus, to overcome the problem imposed by dimension constraints, in the training transmission, the analog matrices at both the transmitter and receiver should be adjusted. The idea is similar to beam scan algorithms. Based on this idea, as shown by Scheme 2 in Fig. 2.1 for channel estimation, the corresponding signal model is expressed by

$$\boldsymbol{Y}_{\mathrm{D}} = \boldsymbol{G}_{\mathrm{A}}\boldsymbol{H}\boldsymbol{F}_{\mathrm{A}}\boldsymbol{X} + \boldsymbol{G}_{\mathrm{A}}\boldsymbol{N}, \tag{2.4}$$

where $\boldsymbol{N}$ is the corresponding noise matrix, and

$$\begin{aligned}\boldsymbol{G}_{\mathrm{A}} &= \left[\boldsymbol{G}_{\mathrm{A},1}^{\mathrm{T}} \cdots \boldsymbol{G}_{\mathrm{A,K}}^{\mathrm{T}}\right]^{\mathrm{T}},\\ \boldsymbol{F}_{\mathrm{A}} &= \left[\boldsymbol{F}_{\mathrm{A},1} \cdots \boldsymbol{F}_{\mathrm{A,K}}\right],\\ \boldsymbol{X} &= \left[\boldsymbol{X}_1^{\mathrm{T}} \cdots \boldsymbol{X}_K^{\mathrm{T}}\right]^{\mathrm{T}}.\end{aligned} \tag{2.5}$$

$\boldsymbol{F}_{\mathrm{A}}$ should be row full rank and $\boldsymbol{G}_{\mathrm{A}}$ should be column full rank. The composite noise term $\boldsymbol{N}$ in (2.4) also satisfies the following Kronecker product structure [31]

$$\boldsymbol{N} = \boldsymbol{\Sigma}_{\mathrm{N}}^{\frac{1}{2}}\boldsymbol{N}_{\mathrm{W}}\boldsymbol{\Psi}_{\mathrm{N}}^{\frac{1}{2}}, \tag{2.6}$$

where each element of $\boldsymbol{N}_{\mathrm{W}}$ is an i.i.d. Gaussian random variable with zero mean and unit variance. It is worth highlighting that within the coherence time, the channel matrix is considered to be time-invariant. Based on (2.4), the estimation of the channel matrix $\boldsymbol{H}$ becomes achievable.

The remaining task of is to recover $\boldsymbol{H}$ from $\boldsymbol{Y}_{\mathrm{D}}$. Following the idea in traditional full digital MIMO channel estimation, the signal model (2.4) may first be transferred into a vector form

$$\mathrm{vec}(\boldsymbol{Y}_{\mathrm{D}}) = \left(\left(\boldsymbol{X}^{\mathrm{T}}\boldsymbol{F}_{\mathrm{A}}^{\mathrm{T}}\right) \otimes \boldsymbol{G}_{\mathrm{A}}\right)\mathrm{vec}(\boldsymbol{H}) + \left(\boldsymbol{I} \otimes \boldsymbol{G}_{\mathrm{A}}\right)\mathrm{vec}(\boldsymbol{N}). \tag{2.7}$$

In the full digital MIMO case without the analog matrices $\boldsymbol{F}_{\mathrm{A}}$ and $\boldsymbol{G}_{\mathrm{A}}$, this model is of course extensively used. In hybrid MIMO, however, the existence of the analog matrices $\boldsymbol{F}_{\mathrm{A}}$ and $\boldsymbol{G}_{\mathrm{A}}$ makes the channel estimation more challenging.

Moreover, when $\boldsymbol{H}$ is a very large matrix, the widely used channel estimators, such as the LS and LMMSE estimators, suffer from prohibitively high complexity. To see why vectorization-based channel estimation is not preferred in practical implementations, we examine its complexity. It is well-known that the computation complexity of matrix inverse is on the order of N^3, denoted as $\mathcal{O}(N^3)$, where N

is the matrix dimension. Then for a channel matrix of dimension 128×128, the complexity of the LS or LMMSE channel estimator is $\mathcal{O}\big((128 \times 128)^3\big)$, which is excessive at the time of writing.

Therefore, the key task of training optimization for hybrid MIMO systems is how to design a low-complexity channel estimator under various physical constraints, such as power constraints at the transmitter and the constant modulus constraints on the analog matrices, while maintaining excellent estimation accuracy. This motivates our work.

Remark 1 We would like to point out that deep learning (DL) algorithms can also be chosen for estimating the channel matrix of the hybrid massive MIMO systems. This approach is different from our signal processing-based perspective. Generally speaking, DL algorithms view the estimation procedure as a black box, while signal processing techniques usually aim to reveal some physical insights. DL algorithms are more suited to problems having no signal models.

2.3 Left-Right Channel Estimation

Based on the signal model (2.4), in order to control the dimension of the channel estimator, the following channel estimator is proposed, in which the channel matrix is estimated as

$$\widehat{\boldsymbol{H}} = \boldsymbol{G}_{\mathrm{D,L}} \boldsymbol{Y}_{\mathrm{D}} \boldsymbol{G}_{\mathrm{D,R}}, \tag{2.8}$$

where $\boldsymbol{G}_{\mathrm{D,L}}$ and $\boldsymbol{G}_{\mathrm{D,R}}$ are the left and right channel estimators, respectively. Different from the traditional vectorization-based channel estimator, here the high-dimensional channel estimator is replaced by two low-dimensional channel estimators, and the corresponding channel estimation complexity is significantly reduced. We now investigate in depth how to find the left and right channel estimators in both scenarios with and without channel statistical information.

2.3.1 *Without Channel Statistics*

We may recover $\boldsymbol{H}$ using the idea of 'zero-forcing', that is, using the following $\boldsymbol{G}_{\mathrm{D,L}}$ and $\boldsymbol{G}_{\mathrm{D,R}}$

$$\begin{aligned} \boldsymbol{G}_{\mathrm{D,L}} &= \big(\boldsymbol{G}_{\mathrm{A}}^{\mathrm{H}} \boldsymbol{G}_{\mathrm{A}}\big)^{-1} \boldsymbol{G}_{\mathrm{A}}^{\mathrm{H}}, \\ \boldsymbol{G}_{\mathrm{D,R}} &= \boldsymbol{X}^{\mathrm{H}} \boldsymbol{F}_{\mathrm{A}}^{\mathrm{H}} \big(\boldsymbol{F}_{\mathrm{A}} \boldsymbol{X} \boldsymbol{X}^{\mathrm{H}} \boldsymbol{F}_{\mathrm{A}}^{\mathrm{H}}\big)^{-1}. \end{aligned} \tag{2.9}$$

Clearly, if the following two equalities hold

$$\boldsymbol{G}_{\mathrm{A}}^{\mathrm{H}} \boldsymbol{G}_{\mathrm{A}} \propto \boldsymbol{I}, \quad \boldsymbol{F}_{\mathrm{A}} \boldsymbol{X} \boldsymbol{X}^{\mathrm{H}} \boldsymbol{F}_{\mathrm{A}}^{\mathrm{H}} \propto \boldsymbol{I}, \tag{2.10}$$

then channel estimation is greatly simplified. By using the white sequence, we have $\boldsymbol{X}\boldsymbol{X}^{\mathrm{H}} \propto \boldsymbol{I}$, and (2.10) becomes

$$\boldsymbol{F}_{\mathrm{A}}\boldsymbol{F}_{\mathrm{A}}^{\mathrm{H}} \propto \boldsymbol{I}. \tag{2.11}$$

Therefore, we can design the analog matrices $\boldsymbol{F}_{\mathrm{A}}$ and $\boldsymbol{G}_{\mathrm{A}}$ appropriately to achieve (2.11). Each element of $\boldsymbol{G}_{\mathrm{A}}$ and $\boldsymbol{F}_{\mathrm{A}}$ has constant modulus. To satisfy these constraints, the most natural choice is to construct $\boldsymbol{G}_{\mathrm{A}}$ and $\boldsymbol{F}_{\mathrm{A}}$ based on DFT matrix.

It can be seen that for large-scale MIMO channel estimation without any channel statistic information, the traditional white sequence is still an effective candidate. It is worth highlighting that this 'zero-forcing' channel estimation algorithm suffers from a serious performance loss in the low SNR regime. Specifically, the noise power may be significantly enhanced by this channel estimator [32].

2.3.2 LMMSE Channel Estimator

Given the channel statistics, specifically, the second-order statistics of the channel matrix and the noise covariance matrix, the LMMSE estimator generally offers much better estimation accuracy than the previous 'zero-forcing' design. Hence we focus on the LMMSE channel estimation for hybrid MIMO systems with the aid of channel statistics. It is worth emphasizing that because of $\boldsymbol{G}_{\mathrm{A}}$, it is challenging to directly estimate $\boldsymbol{H}$ in a linear form. In our work, we adopt two steps to realize channel estimation. At the first step, $\boldsymbol{G}_{\mathrm{D,L}}\boldsymbol{G}_{\mathrm{A}}\boldsymbol{H}$ is estimated instead of $\boldsymbol{H}$. The corresponding estimation MSE matrix is derived to be (2.12).

$$\begin{aligned}
&\boldsymbol{\Phi}_{\mathrm{MSE}}(\boldsymbol{F}_{\mathrm{A}}, \boldsymbol{X}, \boldsymbol{G}_{\mathrm{A}}, \boldsymbol{G}_{\mathrm{D,L}}, \boldsymbol{G}_{\mathrm{D,R}}) \\
=&\mathbb{E}\big\{(\boldsymbol{G}_{\mathrm{D,L}}\boldsymbol{Y}_{\mathrm{D}}\boldsymbol{G}_{\mathrm{D,R}} - \boldsymbol{G}_{\mathrm{D,L}}\boldsymbol{G}_{\mathrm{A}}\boldsymbol{H})^{\mathrm{H}}(\boldsymbol{G}_{\mathrm{D,L}}\boldsymbol{Y}_{\mathrm{D}}\boldsymbol{G}_{\mathrm{D,R}} - \boldsymbol{G}_{\mathrm{D,L}}\boldsymbol{G}_{\mathrm{A}}\boldsymbol{H})\big\} \\
=&(\boldsymbol{F}_{\mathrm{A}}\boldsymbol{X}\boldsymbol{G}_{\mathrm{D,R}} - \boldsymbol{I})^{\mathrm{H}}\boldsymbol{\Psi}_{\mathrm{H}}\mathrm{Tr}\big(\boldsymbol{\Sigma}_{\mathrm{H}}\boldsymbol{G}_{\mathrm{A}}^{\mathrm{H}}\boldsymbol{G}_{\mathrm{D,L}}^{\mathrm{H}}\boldsymbol{G}_{\mathrm{D,L}}\boldsymbol{G}_{\mathrm{A}}\big)(\boldsymbol{F}_{\mathrm{A}}\boldsymbol{X}\boldsymbol{G}_{\mathrm{D,R}} - \boldsymbol{I}) \\
&+ \boldsymbol{G}_{\mathrm{D,R}}^{\mathrm{H}}\boldsymbol{\Psi}_{\mathrm{N}}\boldsymbol{G}_{\mathrm{D,R}}\mathrm{Tr}\big(\boldsymbol{G}_{\mathrm{A}}^{\mathrm{H}}\boldsymbol{G}_{\mathrm{D,L}}^{\mathrm{H}}\boldsymbol{G}_{\mathrm{D,L}}\boldsymbol{G}_{\mathrm{A}}\boldsymbol{\Sigma}_{\mathrm{N}}\big).
\end{aligned} \tag{2.12}$$

In the signal model, the training sequence is on the right hand side of the channel matrix and hence the right digital channel estimator will play a more important role in the channel estimation. Note that there is no constraint imposed on the digital estimator $\boldsymbol{G}_{\mathrm{D,R}}$ at the destination. The sum MSE Tr $\big(\boldsymbol{\Phi}_{\mathrm{MSE}}(\boldsymbol{F}_{\mathrm{A}}, \boldsymbol{X}, \boldsymbol{G}_{\mathrm{A}}, \boldsymbol{G}_{\mathrm{D,L}}, \boldsymbol{G}_{\mathrm{D,R}})\big)$ is a quadratic function with respect to $\boldsymbol{G}_{\mathrm{D,R}}$. Then based on complex matrix derivatives the optimal $\boldsymbol{G}_{\mathrm{D,R}}$ can be derived in closed-form as

$$\begin{aligned}
\boldsymbol{G}_{\mathrm{D,R}}^{\mathrm{opt}} =&\Big(\mathrm{Tr}\big(\boldsymbol{\Sigma}_{\mathrm{H}}\boldsymbol{G}_{\mathrm{A}}^{\mathrm{H}}\boldsymbol{G}_{\mathrm{D,L}}^{\mathrm{H}}\boldsymbol{G}_{\mathrm{D,L}}\boldsymbol{G}_{\mathrm{A}}\big)\boldsymbol{X}^{\mathrm{H}}\boldsymbol{F}_{\mathrm{A}}^{\mathrm{H}}\boldsymbol{\Psi}_{\mathrm{H}}\boldsymbol{F}_{\mathrm{A}}\boldsymbol{X} \\
&+ \boldsymbol{\Psi}_{\mathrm{N}}\mathrm{Tr}\big(\boldsymbol{G}_{\mathrm{A}}^{\mathrm{H}}\boldsymbol{G}_{\mathrm{D,L}}^{\mathrm{H}}\boldsymbol{G}_{\mathrm{D,L}}\boldsymbol{G}_{\mathrm{A}}\boldsymbol{\Sigma}_{\mathrm{N}}\big)\Big)^{-1}\boldsymbol{X}^{\mathrm{H}}\boldsymbol{F}_{\mathrm{A}}^{\mathrm{H}}\boldsymbol{\Psi}_{\mathrm{H}} \\
&\times \mathrm{Tr}\big(\boldsymbol{\Sigma}_{\mathrm{H}}\boldsymbol{G}_{\mathrm{A}}^{\mathrm{H}}\boldsymbol{G}_{\mathrm{D,L}}^{\mathrm{H}}\boldsymbol{G}_{\mathrm{D,L}}\boldsymbol{G}_{\mathrm{A}}\big).
\end{aligned} \tag{2.13}$$

Based on the properties of positive semi-definite matrices, the following matrix inequality holds for the right digital channel estimator [32]

$$\boldsymbol{\Phi}_{\text{MSE}}\big(\boldsymbol{F}_{\text{A}}, \boldsymbol{X}, \boldsymbol{G}_{\text{A}}, \boldsymbol{G}_{\text{D,L}}, \boldsymbol{G}_{\text{D,R}}^{\text{opt}}\big) \preceq \boldsymbol{\Phi}_{\text{MSE}}\big(\boldsymbol{F}_{\text{A}}, \boldsymbol{X}, \boldsymbol{G}_{\text{A}}, \boldsymbol{G}_{\text{D,L}}, \boldsymbol{G}_{\text{D,R}}\big). \tag{2.14}$$

In other words, the optimal right digital channel estimator minimizes the channel estimation MSE matrix in the positive semi-definite matrix domain.

Substituting $\boldsymbol{G}_{\text{D,R}}^{\text{opt}}$ into (2.12), the channel estimation MSE matrix (2.12) is reformulated as (2.15).

$$\begin{aligned}
&\boldsymbol{\Phi}_{\text{MSE}}(\boldsymbol{F}_{\text{A}}, \boldsymbol{X}, \boldsymbol{G}_{\text{A}}, \boldsymbol{G}_{\text{D,L}}, \boldsymbol{G}_{\text{D,R}}^{\text{opt}})\\
=&\boldsymbol{\Psi}_{\text{H}}\text{Tr}(\boldsymbol{\Sigma}_{\text{H}}\boldsymbol{G}_{\text{A}}^{\text{H}}\boldsymbol{G}_{\text{D,L}}^{\text{H}}\boldsymbol{G}_{\text{D,L}}\boldsymbol{G}_{\text{A}}) - \boldsymbol{\Psi}_{\text{H}}\text{Tr}(\boldsymbol{\Sigma}_{\text{H}}\boldsymbol{G}_{\text{A}}^{\text{H}}\boldsymbol{G}_{\text{D,L}}^{\text{H}}\boldsymbol{G}_{\text{D,L}}\boldsymbol{G}_{\text{A}})\boldsymbol{F}_{\text{A}}\boldsymbol{X}\\
&\times\Big(\text{Tr}(\boldsymbol{\Sigma}_{\text{H}}\boldsymbol{G}_{\text{A}}^{\text{H}}\boldsymbol{G}_{\text{D,L}}^{\text{H}}\boldsymbol{G}_{\text{D,L}}\boldsymbol{G}_{\text{A}})\boldsymbol{X}^{\text{H}}\boldsymbol{F}_{\text{A}}^{\text{H}}\boldsymbol{\Psi}_{\text{H}}\boldsymbol{F}_{\text{A}}\boldsymbol{X} + \boldsymbol{\Psi}_{\text{N}}\text{Tr}(\boldsymbol{G}_{\text{A}}^{\text{H}}\boldsymbol{G}_{\text{D,L}}^{\text{H}}\boldsymbol{G}_{\text{D,L}}\boldsymbol{G}_{\text{A}}\boldsymbol{\Sigma}_{\text{N}})\Big)^{-1}\\
&\times\boldsymbol{X}^{\text{H}}\boldsymbol{F}_{\text{A}}^{\text{H}}\boldsymbol{\Psi}_{\text{H}}\text{Tr}(\boldsymbol{\Sigma}_{\text{H}}\boldsymbol{G}_{\text{A}}^{\text{H}}\boldsymbol{G}_{\text{D,L}}^{\text{H}}\boldsymbol{G}_{\text{D,L}}\boldsymbol{G}_{\text{A}})\\
=&\text{Tr}(\boldsymbol{\Sigma}_{\text{H}}\boldsymbol{G}_{\text{A}}^{\text{H}}\boldsymbol{G}_{\text{D,L}}^{\text{H}}\boldsymbol{G}_{\text{D,L}}\boldsymbol{G}_{\text{A}})\left(\boldsymbol{\Psi}_{\text{H}}^{-1} + \frac{\text{Tr}(\boldsymbol{\Sigma}_{\text{H}}\boldsymbol{G}_{\text{A}}^{\text{H}}\boldsymbol{G}_{\text{D,L}}^{\text{H}}\boldsymbol{G}_{\text{D,L}}\boldsymbol{G}_{\text{A}})\boldsymbol{F}_{\text{A}}\boldsymbol{X}\boldsymbol{\Psi}_{\text{N}}^{-1}\boldsymbol{X}^{\text{H}}\boldsymbol{F}_{\text{A}}^{\text{H}}}{\text{Tr}(\boldsymbol{\Sigma}_{\text{N}}\boldsymbol{G}_{\text{A}}^{\text{H}}\boldsymbol{G}_{\text{D,L}}^{\text{H}}\boldsymbol{G}_{\text{D,L}}\boldsymbol{G}_{\text{A}})}\right)^{-1}.
\end{aligned} \tag{2.15}$$

It is worth highlighting that from (2.15), the term $\text{Tr}\big(\boldsymbol{\Sigma}_{\text{H}}\boldsymbol{G}_{\text{A}}^{\text{H}}\boldsymbol{G}_{\text{D,L}}^{\text{H}}\boldsymbol{G}_{\text{D,L}}\boldsymbol{G}_{\text{A}}\big)$ can be interpreted as a scaling factor which does not affect the system performance. If the scaling factor is taken into account in optimization, there will be a trivial conclusion that the optimal $\boldsymbol{G}_{\text{D,L}}$ should be an all-zero matrix as in this case the channel estimation MSE will be zero. This is definitely wrong because when $\boldsymbol{G}_{\text{D,L}}$ is zero, the task for $\boldsymbol{G}_{\text{D,R}}$ becomes to estimate a channel matrix with zero covariance matrix. In this case, the corresponding MSE is zero but this result is meaningless. If we focus on the normalized MSE, this term will be removed directly.

The optimal $\boldsymbol{G}_{\text{A}}$ and $\boldsymbol{G}_{\text{D,L}}$ should minimize the following cost function to ensure that the estimation is accurate

$$\min\ \mathbb{E}\big\{\|\boldsymbol{G}_{\text{D,L}}\boldsymbol{G}_{\text{A}}\boldsymbol{H} - \boldsymbol{H}\|_F^2\big\}, \tag{2.16}$$

based on which the optimal $\boldsymbol{G}_{\text{D,L}}$ is given by

$$\boldsymbol{G}_{\text{D,L}}^{\text{opt}} = (\boldsymbol{G}_{\text{A}}^{\text{H}}\boldsymbol{G}_{\text{A}})^{-1}\boldsymbol{G}_{\text{A}}^{\text{H}}. \tag{2.17}$$

In order to guarantee the channel estimation is feasible the analog estimator $\boldsymbol{G}_{\text{A}}$ should be column full rank. The simplest method is to choose corresponding columns of the DFT matrix $\boldsymbol{U}_{\text{DFE}}$ to construct $\boldsymbol{G}_{\text{A}}$. Therefore, the optimal analog equalizer at the destination is

$$\boldsymbol{G}_{\text{A}} = [\boldsymbol{U}_{\text{DFT}}]_{:,N}, \tag{2.18}$$

where N is the number of data streams. In addition, $\boldsymbol{G}_{\text{D,L}}$ equals to the left inverse of $\boldsymbol{G}_{\text{A}}$, i.e.,

$$\boldsymbol{G}_{\mathrm{D,L}} = [\boldsymbol{U}_{\mathrm{DFT}}]^{\mathrm{H}}_{:,N}. \tag{2.19}$$

Based on the results given above, a simplified channel estimation MSE matrix is defined as

$$\begin{aligned}&\boldsymbol{\Phi}_{\mathrm{MSE}}(\boldsymbol{F}_{\mathrm{A}}, \boldsymbol{X}, \boldsymbol{G}_{\mathrm{A}}, \boldsymbol{G}^{\mathrm{opt}}_{\mathrm{D,L}}, \boldsymbol{G}^{\mathrm{opt}}_{\mathrm{D,R}})\\ =&\beta_1 \left(\boldsymbol{\Psi}_{\mathrm{H}}^{-1} + \alpha_1 \boldsymbol{F}_{\mathrm{A}}\boldsymbol{X}\boldsymbol{\Psi}_{\mathrm{N}}^{-1}\boldsymbol{X}^{\mathrm{H}}\boldsymbol{F}_{\mathrm{A}}^{\mathrm{H}}\right)^{-1} \triangleq \boldsymbol{\Phi}_{\mathrm{MSE}}(\boldsymbol{F}_{\mathrm{A}}, \boldsymbol{X}),\end{aligned} \tag{2.20}$$

where β_1 and α_1 are defined respectively as

$$\begin{aligned}\beta_1 =& \mathrm{Tr}\big(\boldsymbol{\Sigma}_{\mathrm{H}}\boldsymbol{G}_{\mathrm{A}}^{\mathrm{H}}\boldsymbol{G}_{\mathrm{D,L}}^{\mathrm{H}}\boldsymbol{G}_{\mathrm{D,L}}\boldsymbol{G}_{\mathrm{A}}\big) = \mathrm{Tr}\big(\boldsymbol{\Sigma}_{\mathrm{H}}\big),\\ \alpha_1 =& \frac{\mathrm{Tr}\big(\boldsymbol{\Sigma}_{\mathrm{H}}\boldsymbol{G}_{\mathrm{A}}^{\mathrm{H}}\boldsymbol{G}_{\mathrm{D,L}}^{\mathrm{H}}\boldsymbol{G}_{\mathrm{D,L}}\boldsymbol{G}_{\mathrm{A}}\big)}{\mathrm{Tr}\big(\boldsymbol{\Sigma}_{\mathrm{N}}\boldsymbol{G}_{\mathrm{A}}^{\mathrm{H}}\boldsymbol{G}_{\mathrm{D,L}}^{\mathrm{H}}\boldsymbol{G}_{\mathrm{D,L}}\boldsymbol{G}_{\mathrm{A}}\big)} = \frac{\mathrm{Tr}\big(\boldsymbol{\Sigma}_{\mathrm{H}}\big)}{\mathrm{Tr}\big(\boldsymbol{\Sigma}_{\mathrm{N}}\big)}.\end{aligned} \tag{2.21}$$

Generally speaking, the training optimization for hybrid MIMO systems aims at optimizing a matrix monotonic function with respect to $\boldsymbol{\Phi}_{\mathrm{MSE}}(\boldsymbol{F}_{\mathrm{A}}, \boldsymbol{X})$ [30], formulated as

$$\begin{aligned}&\min_{\boldsymbol{F}_{\mathrm{A}},\boldsymbol{X}} \ f\big(\boldsymbol{\Phi}_{\mathrm{MSE}}(\boldsymbol{F}_{\mathrm{A}}, \boldsymbol{X})\big)\\ &\ \text{s.t.}\ \mathrm{Tr}\big(\boldsymbol{F}_{\mathrm{A}}\boldsymbol{X}\boldsymbol{X}^{\mathrm{H}}\boldsymbol{F}_{\mathrm{A}}^{\mathrm{H}}\big) \le P,\ \boldsymbol{F}_{\mathrm{A}} \in \mathcal{F},\end{aligned} \tag{2.22}$$

where $f(\cdot)$ is a matrix monotonically increasing function with respect to $\boldsymbol{\Phi}_{\mathrm{MSE}}(\boldsymbol{F}_{\mathrm{A}}, \boldsymbol{X})$, and P is the maximum transmit power constraint for the training design, while the set $\mathcal{F}$ denotes the analog matrix set with proper dimensions, in which each element of a analog matrix has constant modulus.

In the following, we investigate in depth the two training optimizations with specific performance metrics, namely, the MSE minimization and MI maximization.

2.3.3 Sum MSE Minimization

First, we focus on sum MSE minimization and the corresponding objective is the sum of the MSE of each channel matrix element. In other words, it is the sum of the diagonal elements of the MSE matrix. For the sum MSE minimization, the training optimization problem is equivalent to

$$\begin{aligned}&\min_{\boldsymbol{F}_{\mathrm{A}},\boldsymbol{X}} \ \mathrm{Tr}\left(\left(\boldsymbol{\Psi}_{\mathrm{H}}^{-1} + \alpha_1 \boldsymbol{F}_{\mathrm{A}}\boldsymbol{X}\boldsymbol{\Psi}_{\mathrm{N}}^{-1}\boldsymbol{X}^{\mathrm{H}}\boldsymbol{F}_{\mathrm{A}}^{\mathrm{H}}\right)^{-1}\right)\\ &\ \text{s.t.}\ \mathrm{Tr}\big(\boldsymbol{F}_{\mathrm{A}}\boldsymbol{X}\boldsymbol{X}^{\mathrm{H}}\boldsymbol{F}_{\mathrm{A}}^{\mathrm{H}}\big) \le P,\ \boldsymbol{F}_{\mathrm{A}} \in \mathcal{F},\end{aligned} \tag{2.23}$$

where $\boldsymbol{F}_{\mathrm{A}}$ is the analog matrix at the source, which has row full rank. In other words, $\boldsymbol{F}_{\mathrm{A}}\boldsymbol{F}_{\mathrm{A}}^{\mathrm{H}}$ is full rank.

Defining the following matrix variable

$$\widetilde{\boldsymbol{X}} = \big(\boldsymbol{F}_{\mathrm{A}}\boldsymbol{F}_{\mathrm{A}}^{\mathrm{H}}\big)^{\frac{1}{2}}\boldsymbol{X}, \tag{2.24}$$

the optimization problem (2.23) is equivalent to the optimization problem (2.25).

$$\begin{aligned}&\min_{\boldsymbol{F}_{\mathrm{A}},\boldsymbol{X}} \mathrm{Tr}\left(\left(\boldsymbol{\Psi}_{\mathrm{H}}^{-1}+\alpha_1\boldsymbol{F}_{\mathrm{A}}(\boldsymbol{F}_{\mathrm{A}}\boldsymbol{F}_{\mathrm{A}}^{\mathrm{H}})^{-\frac{1}{2}}\widetilde{\boldsymbol{X}}\boldsymbol{\Psi}_{\mathrm{N}}^{-1}\widetilde{\boldsymbol{X}}^{\mathrm{H}}(\boldsymbol{F}_{\mathrm{A}}\boldsymbol{F}_{\mathrm{A}}^{\mathrm{H}})^{-\frac{1}{2}}\boldsymbol{F}_{\mathrm{A}}^{\mathrm{H}}\right)^{-1}\right)\\&\mathrm{s.t.}\,\mathrm{Tr}\left(\boldsymbol{F}_{\mathrm{A}}(\boldsymbol{F}_{\mathrm{A}}\boldsymbol{F}_{\mathrm{A}}^{\mathrm{H}})^{-\frac{1}{2}}\widetilde{\boldsymbol{X}}\widetilde{\boldsymbol{X}}^{\mathrm{H}}(\boldsymbol{F}_{\mathrm{A}}\boldsymbol{F}_{\mathrm{A}}^{\mathrm{H}})^{-\frac{1}{2}}\boldsymbol{F}_{\mathrm{A}}^{\mathrm{H}}\right)\le P,\ \boldsymbol{F}_{\mathrm{A}}\in\mathcal{F}.\end{aligned} \tag{2.25}$$

Given the optimal solution of $\boldsymbol{F}_{\mathrm{A}}$ and based on the following singular value decomposition (SVD)

$$\boldsymbol{F}_{\mathrm{A}}=\boldsymbol{U}_{\boldsymbol{F}_{\mathrm{A}}}\boldsymbol{\Lambda}_{\boldsymbol{F}_{\mathrm{A}}}\boldsymbol{V}_{\boldsymbol{F}_{\mathrm{A}}}^{\mathrm{H}}\ \text{with}\ \boldsymbol{\Lambda}_{\boldsymbol{F}_{\mathrm{A}}}\searrow, \tag{2.26}$$

the optimization problem (2.25) can be transferred into the optimization problem

$$\begin{aligned}&\min_{\boldsymbol{F}_{\mathrm{A}},\boldsymbol{X}}\ \mathrm{Tr}\left(\left(\boldsymbol{\Psi}_{\mathrm{H}}^{-1}+\alpha_1\boldsymbol{U}_{\boldsymbol{F}_{\mathrm{A}}}\left[\boldsymbol{V}_{\boldsymbol{F}_{\mathrm{A}}}^{\mathrm{H}}\widetilde{\boldsymbol{X}}\boldsymbol{\Psi}_{\mathrm{N}}^{-1}\widetilde{\boldsymbol{X}}^{\mathrm{H}}\boldsymbol{V}_{\boldsymbol{F}_{\mathrm{A}}}\right]_{[1:N_R;1:N_R]}\boldsymbol{U}_{\boldsymbol{F}_{\mathrm{A}}}^{\mathrm{H}}\right)^{-1}\right)\\&\mathrm{s.t.}\ \mathrm{Tr}\left(\left[\boldsymbol{V}_{\boldsymbol{F}_{\mathrm{A}}}^{\mathrm{H}}\widetilde{\boldsymbol{X}}\widetilde{\boldsymbol{X}}^{\mathrm{H}}\boldsymbol{V}_{\boldsymbol{F}_{\mathrm{A}}}\right]_{[1:N_R;1:N_R]}\right)\le P,\ \boldsymbol{F}_{\mathrm{A}}\in\mathcal{F},\end{aligned} \tag{2.27}$$

where N_R is the number of rows in $\boldsymbol{F}_{\mathrm{A}}$.

In order to derive the optimal solution, we first quote the following inequality [33]

$$\mathrm{Tr}(\boldsymbol{A}+\boldsymbol{B})^{-1}\ge\sum_i 1/(\lambda_{\boldsymbol{A},i}+\lambda_{\boldsymbol{B},N-i+1}), \tag{2.28}$$

where $\lambda_{\boldsymbol{Z},i}$ denotes the ith largest eigenvalue of $\boldsymbol{Z}$. The equality holds when the unitary matrices of the eigen value decomposition (EVD) of $\boldsymbol{A}$ and $\boldsymbol{B}$ satisfy the following relationship

$$\boldsymbol{U}_{\boldsymbol{A}}=\bar{\boldsymbol{U}}_{\boldsymbol{B}}, \tag{2.29}$$

where the unitary matrices $\boldsymbol{U}_{\boldsymbol{A}}$ and $\bar{\boldsymbol{U}}_{\boldsymbol{B}}$ are defined based on the following EVDs

$$\begin{aligned}\boldsymbol{A}&=\boldsymbol{U}_{\boldsymbol{A}}\boldsymbol{\Lambda}_{\boldsymbol{A}}\boldsymbol{U}_{\boldsymbol{A}}^{\mathrm{H}}\ \text{with}\ \boldsymbol{\Lambda}_{\boldsymbol{A}}\searrow,\\\boldsymbol{B}&=\bar{\boldsymbol{U}}_{\boldsymbol{B}}\bar{\boldsymbol{\Lambda}}_{\boldsymbol{B}}\bar{\boldsymbol{U}}_{\boldsymbol{B}}^{\mathrm{H}}\ \text{with}\ \bar{\boldsymbol{\Lambda}}_{\boldsymbol{B}}\nearrow.\end{aligned} \tag{2.30}$$

From (2.28) and (2.27), the following conclusion can be obtained using the matrix-monotonic optimization framework [30].

Conclution 2.1 *The optimal* $\widetilde{\boldsymbol{X}}$ *for the optimization problem (2.25) satisfies the following optimal structure*

$$\widetilde{\boldsymbol{X}}=\boldsymbol{V}_{\boldsymbol{F}_{\mathrm{A}}}\mathrm{diag}\{\boldsymbol{U}_{\boldsymbol{F}_{\mathrm{A}}}^{\mathrm{H}}\boldsymbol{U}_{\boldsymbol{\Psi}_{\mathrm{H}}}\boldsymbol{\Lambda}_{\boldsymbol{X}},\boldsymbol{0}\}\bar{\boldsymbol{U}}_{\boldsymbol{\Psi}_{\mathrm{N}}}^{\mathrm{H}}, \tag{2.31}$$

where $\boldsymbol{\Lambda}_{\boldsymbol{X}}$ is a diagonal matrix, while the unitary matrices $\boldsymbol{U}_{\boldsymbol{\Psi}_{\mathrm{H}}}$ and $\bar{\boldsymbol{U}}_{\boldsymbol{\Psi}_{\mathrm{N}}}$ are defined based on the following EVDs

$$\boldsymbol{\Psi}_{\mathrm{H}} = \boldsymbol{U}_{\boldsymbol{\Psi}_{\mathrm{H}}}\boldsymbol{\Lambda}_{\boldsymbol{\Psi}_{\mathrm{H}}}\boldsymbol{U}_{\boldsymbol{\Psi}_{\mathrm{H}}}^{\mathrm{H}} \text{ with } \boldsymbol{\Lambda}_{\boldsymbol{\Psi}_{\mathrm{H}}} \searrow, \ \boldsymbol{\Psi}_{\mathrm{N}} = \bar{\boldsymbol{U}}_{\boldsymbol{\Psi}_{\mathrm{N}}}\bar{\boldsymbol{\Lambda}}_{\boldsymbol{\Psi}_{\mathrm{N}}}\bar{\boldsymbol{U}}_{\boldsymbol{\Psi}_{\mathrm{N}}}^{\mathrm{H}} \text{ with } \bar{\boldsymbol{\Lambda}}_{\boldsymbol{\Psi}_{\mathrm{N}}} \nearrow. \tag{2.32}$$

Using (2.31) in (2.25), $\boldsymbol{F}_{\mathrm{A}}$ can be removed, i.e., it does not affect the optimization, and we have the following conclusion.

Conclution 2.2 *For the optimization problem (2.25), the optimal analog matrix $\boldsymbol{F}_{\mathrm{A}}$ is an arbitrary row full rank matrix with constant modulus elements.*

Based on the optimal structure given in Conclusion 2.1, the original sum-MSE minimization problem can be simplified into the following optimization problem

$$\begin{aligned} &\min_{\{f_i^2\}} \sum_i \frac{1}{\frac{1}{\lambda_{\boldsymbol{\Psi}_{\mathrm{H}},i}} + \frac{\alpha_1 f_i^2}{\lambda_{\boldsymbol{\Psi}_{\mathrm{N}},i}}} \\ &\text{s.t.} \sum_i f_i^2 \le P, \end{aligned} \tag{2.33}$$

where $f_i = [\boldsymbol{\Lambda}_{\boldsymbol{X}}]_{i,i}$, $\lambda_{\boldsymbol{\Psi}_{\mathrm{H}},i} = [\boldsymbol{\Lambda}_{\boldsymbol{\Psi}_{\mathrm{H}}}]_{i,i}$, and $\lambda_{\boldsymbol{\Psi}_{\mathrm{N}},i} = [\bar{\boldsymbol{\Lambda}}_{\boldsymbol{\Psi}_{\mathrm{N}}}]_{i,i}$. The optimal solution of (2.33) is a standard water-filling solution given by [34]

$$f_i^2 = \left(\sqrt{\frac{\lambda_{\boldsymbol{\Psi}_{\mathrm{N}},i}}{\alpha_1 \mu}} - \frac{\lambda_{\boldsymbol{\Psi}_{\mathrm{N}},i}}{\alpha_1 \lambda_{\boldsymbol{\Psi}_{\mathrm{H}},i}}\right)^+, \tag{2.34}$$

where μ is the Lagrange multiplier for the power constraint in (2.33).

2.3.4 MI Maximization

The MI between the true channel and the estimated channel is another important performance metric for training optimization [14]. Specifically, the MI maximization-based training optimization for hybrid MIMO communications is equivalent to

$$\begin{aligned} &\max_{\boldsymbol{F}_{\mathrm{A}},\boldsymbol{X}} \log\left|\boldsymbol{\Psi}_{\mathrm{H}}^{-1} + \alpha_1 \boldsymbol{F}_{\mathrm{A}}\boldsymbol{X}\boldsymbol{\Psi}_{\mathrm{N}}^{-1}\boldsymbol{X}^{\mathrm{H}}\boldsymbol{F}_{\mathrm{A}}^{\mathrm{H}}\right| \\ &\text{s.t. } \mathrm{Tr}(\boldsymbol{F}_{\mathrm{A}}\boldsymbol{X}\boldsymbol{X}^{\mathrm{H}}\boldsymbol{F}_{\mathrm{A}}^{\mathrm{H}}) \le P, \ \boldsymbol{F}_{\mathrm{A}} \in \mathcal{F}. \end{aligned} \tag{2.35}$$

Similarly, by defining the following auxiliary variable

$$\widetilde{\boldsymbol{X}} = \left(\boldsymbol{F}_{\mathrm{A}}\boldsymbol{F}_{\mathrm{A}}^{\mathrm{H}}\right)^{\frac{1}{2}}\boldsymbol{X}, \tag{2.36}$$

the training optimization (2.35) is equivalent to (2.37).

$$\begin{aligned}&\max_{\boldsymbol{F}_{\mathrm{A}},\boldsymbol{X}} \log\left|\boldsymbol{\Psi}_{\mathrm{H}}^{-1}+\alpha_1\boldsymbol{F}_{\mathrm{A}}(\boldsymbol{F}_{\mathrm{A}}\boldsymbol{F}_{\mathrm{A}}^{\mathrm{H}})^{-\frac{1}{2}}\widetilde{\boldsymbol{X}}\boldsymbol{\Psi}_{\mathrm{N}}^{-1}\widetilde{\boldsymbol{X}}^{\mathrm{H}}(\boldsymbol{F}_{\mathrm{A}}\boldsymbol{F}_{\mathrm{A}}^{\mathrm{H}})^{-\frac{1}{2}}\boldsymbol{F}_{\mathrm{A}}^{\mathrm{H}}\right|\\&\text{s.t. } \mathrm{Tr}\left(\boldsymbol{F}_{\mathrm{A}}(\boldsymbol{F}_{\mathrm{A}}\boldsymbol{F}_{\mathrm{A}}^{\mathrm{H}})^{-\frac{1}{2}}\widetilde{\boldsymbol{X}}\widetilde{\boldsymbol{X}}^{\mathrm{H}}(\boldsymbol{F}_{\mathrm{A}}\boldsymbol{F}_{\mathrm{A}}^{\mathrm{H}})^{-\frac{1}{2}}\boldsymbol{F}_{\mathrm{A}}^{\mathrm{H}}\right)\leq P,\ \boldsymbol{F}_{\mathrm{A}}\in\mathcal{F}.\end{aligned} \tag{2.37}$$

Similar to the MSE minimization, the optimization (2.37) can be transferred to

$$\begin{aligned}&\max_{\boldsymbol{F}_{\mathrm{A}},\boldsymbol{X}} \log\left|\boldsymbol{\Psi}_{\mathrm{H}}^{-1}+\alpha_1\boldsymbol{U}_{\boldsymbol{F}_{\mathrm{A}}}\left[\boldsymbol{V}_{\boldsymbol{F}_{\mathrm{A}}}^{\mathrm{H}}\widetilde{\boldsymbol{X}}\boldsymbol{\Psi}_{\mathrm{N}}^{-1}\widetilde{\boldsymbol{X}}^{\mathrm{H}}\boldsymbol{V}_{\boldsymbol{F}_{\mathrm{A}}}\right]_{1:N_R,1:N_R}\boldsymbol{U}_{\boldsymbol{F}_{\mathrm{A}}}^{\mathrm{H}}\right|\\&\text{s.t. } \mathrm{Tr}\left(\left[\boldsymbol{V}_{\boldsymbol{F}_{\mathrm{A}}}^{\mathrm{H}}\widetilde{\boldsymbol{X}}\widetilde{\boldsymbol{X}}^{\mathrm{H}}\boldsymbol{V}_{\boldsymbol{F}_{\mathrm{A}}}\right]_{1:N_R,1:N_R}\right)\leq P,\ \boldsymbol{F}_{\mathrm{A}}\in\mathcal{F}.\end{aligned} \tag{2.38}$$

Further consider the inequality [33]

$$\log|\boldsymbol{A}+\boldsymbol{B}|\leq\sum\nolimits_i\log(\lambda_{\boldsymbol{A},i}+\lambda_{\boldsymbol{B},N-i+1}). \tag{2.39}$$

The equality holds when the unitary matrices $\boldsymbol{U}_{\boldsymbol{A}}$ and $\bar{\boldsymbol{U}}_{\boldsymbol{B}}$, defined in (2.30), satisfy $\boldsymbol{U}_{\boldsymbol{A}}=\bar{\boldsymbol{U}}_{\boldsymbol{B}}$. Based on (2.38) and (2.39), we have the following two conclusions using the matrix-monotonic optimization framework [30].

Conclution 2.3 *The optimal $\widetilde{\boldsymbol{X}}$ of the problem (2.37) satisfies the following structure*

$$\widetilde{\boldsymbol{X}}=\boldsymbol{V}_{\boldsymbol{F}_{\mathrm{A}}}\mathrm{diag}\{\boldsymbol{U}_{\boldsymbol{F}_{\mathrm{A}}}^{\mathrm{H}}\boldsymbol{U}_{\boldsymbol{\Psi}_{\mathrm{H}}}\boldsymbol{\Lambda}_{\boldsymbol{X}},\boldsymbol{0}\}\boldsymbol{U}_{\boldsymbol{\Psi}_{\mathrm{N}}}^{\mathrm{H}}. \tag{2.40}$$

Conclution 2.4 *For the optimization problem (2.37), the optimal analog matrix $\boldsymbol{F}_{\mathrm{A}}$ is an arbitrary row full rank matrix with constant modulus elements.*

Based on Conclusion 2.1–2.4, it can be stated that for our training optimization conceived for the sum-MSE and for the MI metrics, the analog matrices and training sequences have the same optimal structure. However, it will be shown later in this section that for the sum-MSE and for the MI metrics, the optimal diagonal matrices in Conclusion 2.1 and 2.3 are different.

Exploiting the optimal structure given in Conclusion 2.3, the MI maximization problem is transferred into the following optimization problem,

$$\begin{aligned}&\max_{\{f_i^2\}}\sum_i\log\left(\frac{1}{\lambda_{\boldsymbol{\Psi}_{\mathrm{H}},i}}+\frac{\alpha_1 f_i^2}{\lambda_{\boldsymbol{\Psi}_{\mathrm{N}},i}}\right)\\&\text{s.t. }\sum_i f_i^2\leq P.\end{aligned} \tag{2.41}$$

The optimal water-filling solution of (2.41) is given by [34]

$$f_i^2=\left(\frac{1}{\mu}-\frac{\lambda_{\boldsymbol{\Psi}_{\mathrm{N}},i}}{\alpha_1\lambda_{\boldsymbol{\Psi}_{\mathrm{H}},i}}\right)^+. \tag{2.42}$$

2.4 Composite Channel Estimation

The idea of composite channel estimation is to estimate $\boldsymbol{G}_{\mathrm{A}}\boldsymbol{H}\boldsymbol{F}_{\mathrm{A}}$ rather than $\boldsymbol{H}$. The advantage is that only a much smaller matrix is to be estimated. For this to work, however, the analog matrices used in channel estimation and data transmission must be the same. Generally, the analog matrices may be different in channel estimation and data transmission. Sometimes, however, the analog matrices are designed based on channel statistics instead of instantaneous CSI [35–37]. Therefore, the analog matrices used in the data transmission are also available in channel estimation. Consequently, in these cases, composite channel estimation is applicable, and the premise of this kind of channel estimation is how to choose the analog matrices $\boldsymbol{G}_{\mathrm{A}}$ and $\boldsymbol{F}_{\mathrm{A}}$.

2.4.1 *Analog Matrix Optimization*

In the data transmission phase, the signal model is

$$\widetilde{\boldsymbol{y}} = \boldsymbol{G}_{\mathrm{A}}\boldsymbol{H}\boldsymbol{F}_{\mathrm{A}}\boldsymbol{F}_{\mathrm{D}}\boldsymbol{x} + \boldsymbol{G}_{\mathrm{A}}\boldsymbol{n}, \tag{2.43}$$

based on which the optimization problem of average capacity maximization is given by

$$\begin{aligned} &\max_{\boldsymbol{G}_{\mathrm{A}},\boldsymbol{F}_{\mathrm{A}},\boldsymbol{F}_{\mathrm{D}}} \ \mathbb{E}\left\{\log\left|\boldsymbol{F}_{\mathrm{D}}^{\mathrm{H}}\boldsymbol{F}_{\mathrm{A}}^{\mathrm{H}}\boldsymbol{H}^{\mathrm{H}}\boldsymbol{G}_{\mathrm{A}}^{\mathrm{H}}\left(\boldsymbol{G}_{\mathrm{A}}\boldsymbol{\Sigma}_{\mathrm{N}}\boldsymbol{G}_{\mathrm{A}}^{\mathrm{H}}\right)^{-1}\boldsymbol{G}_{\mathrm{A}}\boldsymbol{H}\boldsymbol{F}_{\mathrm{A}}\boldsymbol{F}_{\mathrm{D}} + \boldsymbol{I}\right|\right\} \\ &\quad \text{s.t. } \mathrm{Tr}\left(\boldsymbol{F}_{\mathrm{A}}\boldsymbol{F}_{\mathrm{D}}\boldsymbol{F}_{\mathrm{D}}^{\mathrm{H}}\boldsymbol{F}_{\mathrm{A}}^{\mathrm{H}}\right) \le P_{\mathrm{T}}, \end{aligned} \tag{2.44}$$

where P_{T} is the maximum transmit power in data transmission. It is worth highlighting that different from the channel estimation procedure, in data transmission, $\boldsymbol{F}_{\mathrm{A}}$ is column full rank and $\boldsymbol{G}_{\mathrm{A}}$ is row full rank. Based on (2.2) and defining

$$\widetilde{\boldsymbol{F}}_{\mathrm{D}} = \left(\boldsymbol{F}_{\mathrm{A}}^{\mathrm{H}}\boldsymbol{F}_{\mathrm{A}}\right)^{\frac{1}{2}}\boldsymbol{F}_{\mathrm{D}}, \tag{2.45}$$

the average capacity maximization (2.44) can be rewritten as

$$\begin{aligned} \max_{\boldsymbol{G}_{\mathrm{A}},\boldsymbol{F}_{\mathrm{A}},\widetilde{\boldsymbol{F}}_{\mathrm{D}}} \ & \mathbb{E}\Big\{\log\Big|\widetilde{\boldsymbol{F}}_{\mathrm{D}}^{\mathrm{H}}\left(\boldsymbol{F}_{\mathrm{A}}^{\mathrm{H}}\boldsymbol{F}_{\mathrm{A}}\right)^{-\frac{1}{2}}\boldsymbol{F}_{\mathrm{A}}^{\mathrm{H}}\boldsymbol{\Psi}_{\mathrm{H}}^{\frac{1}{2}}\boldsymbol{H}_{\mathrm{W}}^{\mathrm{H}}\boldsymbol{\Sigma}_{\mathrm{H}}^{\frac{1}{2}}\boldsymbol{G}_{\mathrm{A}}^{\mathrm{H}}\left(\boldsymbol{G}_{\mathrm{A}}\boldsymbol{\Sigma}_{\mathrm{N}}\boldsymbol{G}_{\mathrm{A}}^{\mathrm{H}}\right)^{-1} \\ & \quad \times\boldsymbol{G}_{\mathrm{A}}\boldsymbol{\Sigma}_{\mathrm{H}}^{\frac{1}{2}}\boldsymbol{H}_{\mathrm{W}}\boldsymbol{\Psi}_{\mathrm{H}}^{\frac{1}{2}}\boldsymbol{F}_{\mathrm{A}}\left(\boldsymbol{F}_{\mathrm{A}}^{\mathrm{H}}\boldsymbol{F}_{\mathrm{A}}\right)^{-\frac{1}{2}}\widetilde{\boldsymbol{F}}_{\mathrm{D}} + \boldsymbol{I}\Big|\Big\}, \\ \text{s.t. } & \mathrm{Tr}\left(\widetilde{\boldsymbol{F}}_{\mathrm{D}}\widetilde{\boldsymbol{F}}_{\mathrm{D}}^{\mathrm{H}}\right) \le P_{\mathrm{T}}. \end{aligned} \tag{2.46}$$

As proved in Appendix 1, the optimal $\boldsymbol{G}_{\mathrm{A}}$ is the optimal solution of the following optimization problem

$$\max_{\boldsymbol{G}_{\mathrm{A}}} \boldsymbol{\lambda}\left(\boldsymbol{\Sigma}_{\mathrm{H}}^{\frac{1}{2}}\boldsymbol{G}_{\mathrm{A}}^{\mathrm{H}}\left(\boldsymbol{G}_{\mathrm{A}}\boldsymbol{\Sigma}_{\mathrm{N}}\boldsymbol{G}_{\mathrm{A}}^{\mathrm{H}}\right)^{-1}\boldsymbol{G}_{\mathrm{A}}\boldsymbol{\Sigma}_{\mathrm{H}}^{\frac{1}{2}}\right), \tag{2.47}$$

where $\boldsymbol{\lambda}(\boldsymbol{Z}) = \left[\lambda_{Z,1} \cdots \lambda_{Z,N}\right]^{\mathrm{T}}$, which is equivalent to the following optimization problem

$$\max_{\boldsymbol{G}_{\mathrm{A}}} \boldsymbol{\lambda}\left(\left(\boldsymbol{G}_{\mathrm{A}}\boldsymbol{\Sigma}_{\mathrm{N}}\boldsymbol{G}_{\mathrm{A}}^{\mathrm{H}}\right)^{-1/2}\boldsymbol{G}_{\mathrm{A}}\boldsymbol{\Sigma}_{\mathrm{H}}\boldsymbol{G}_{\mathrm{A}}^{\mathrm{H}}\left(\boldsymbol{G}_{\mathrm{A}}\boldsymbol{\Sigma}_{\mathrm{N}}\boldsymbol{G}_{\mathrm{A}}^{\mathrm{H}}\right)^{-1/2}\right). \tag{2.48}$$

Moreover, the optimization problem (2.48) can further be rewritten as

$$\max_{\boldsymbol{G}_{\mathrm{A}}} \boldsymbol{\lambda}\left(\left(\boldsymbol{G}_{\mathrm{A}}\boldsymbol{\Sigma}_{\mathrm{N}}\boldsymbol{G}_{\mathrm{A}}^{\mathrm{H}}\right)^{-1/2}\boldsymbol{G}_{\mathrm{A}}\boldsymbol{\Sigma}_{\mathrm{N}}^{\frac{1}{2}}\boldsymbol{\Sigma}_{\mathrm{N}}^{-\frac{1}{2}}\boldsymbol{\Sigma}_{\mathrm{H}}\boldsymbol{\Sigma}_{\mathrm{N}}^{-\frac{1}{2}}\boldsymbol{\Sigma}_{\mathrm{N}}^{\frac{1}{2}}\boldsymbol{G}_{\mathrm{A}}^{\mathrm{H}}\left(\boldsymbol{G}_{\mathrm{A}}\boldsymbol{\Sigma}_{\mathrm{N}}\boldsymbol{G}_{\mathrm{A}}^{\mathrm{H}}\right)^{-1/2}\right). \tag{2.49}$$

It's worth noting that nonzero singular values of the matrix term $\boldsymbol{\Sigma}_{\mathrm{N}}^{\frac{1}{2}}\boldsymbol{G}_{\mathrm{A}}^{\mathrm{H}}\left(\boldsymbol{G}_{\mathrm{A}}\boldsymbol{\Sigma}_{\mathrm{N}}\boldsymbol{G}_{\mathrm{A}}^{\mathrm{H}}\right)^{\frac{1}{2}}$ are all ones. Then, based on the SVD it may be concluded that in (2.49) the term $\boldsymbol{\Sigma}_{\mathrm{N}}^{\frac{1}{2}}\boldsymbol{G}_{\mathrm{A}}^{\mathrm{H}}\left(\boldsymbol{G}_{\mathrm{A}}\boldsymbol{\Sigma}_{\mathrm{N}}\boldsymbol{G}_{\mathrm{A}}^{\mathrm{H}}\right)^{\frac{1}{2}}$ aims for selecting the first N maximum eigenvalues of the inner matrix term $\boldsymbol{\Sigma}_{\mathrm{N}}^{-\frac{1}{2}}\boldsymbol{\Sigma}_{\mathrm{H}}\boldsymbol{\Sigma}_{\mathrm{N}}^{-\frac{1}{2}}$. The solution of (2.48) is computed by solving the problem

$$\begin{aligned} \min_{\boldsymbol{\Lambda},\boldsymbol{Q},\boldsymbol{G}_{\mathrm{A}}} &\left\|[\boldsymbol{U}_{\mathrm{NHN}}]_{:,N_r}\boldsymbol{\Lambda}\boldsymbol{Q} - \boldsymbol{\Sigma}_{\mathrm{N}}^{\frac{1}{2}}\boldsymbol{G}_{\mathrm{A}}^{\mathrm{H}}\right\|_F^2 \\ \text{s.t. } &\boldsymbol{Q}\boldsymbol{Q}^{\mathrm{H}} = \boldsymbol{I},\ \boldsymbol{G}_{\mathrm{A}} \in \mathcal{F}, \end{aligned} \tag{2.50}$$

which can be transferred into

$$\begin{aligned} \min_{\boldsymbol{\Lambda},\boldsymbol{Q},\boldsymbol{G}_{\mathrm{A}}} &\left\|\boldsymbol{\Sigma}_{\mathrm{N}}^{-\frac{1}{2}}[\boldsymbol{U}_{\mathrm{NHN}}]_{:,N_r}\boldsymbol{\Lambda}\boldsymbol{Q} - \boldsymbol{G}_{\mathrm{A}}^{\mathrm{H}}\right\|_F^2 \\ \text{s.t. } &\boldsymbol{Q}\boldsymbol{Q}^{\mathrm{H}} = \boldsymbol{I},\ \boldsymbol{G}_{\mathrm{A}} \in \mathcal{F}, \end{aligned} \tag{2.51}$$

where the unitary matrix $\boldsymbol{U}_{\mathrm{NHN}}$ is defined in the EVD

$$\boldsymbol{\Sigma}_{\mathrm{N}}^{-\frac{1}{2}}\boldsymbol{\Sigma}_{\mathrm{H}}\boldsymbol{\Sigma}_{\mathrm{N}}^{-\frac{1}{2}} = \boldsymbol{U}_{\mathrm{NHN}}\boldsymbol{\Lambda}_{\mathrm{NHN}}\boldsymbol{U}_{\mathrm{NHN}}^{\mathrm{H}}. \tag{2.52}$$

The problem (2.51) can be solved by alternatively optimizing $\boldsymbol{\Lambda}$, $\boldsymbol{Q}$, and $\boldsymbol{G}_{\mathrm{A}}$. When $\boldsymbol{Q}$ and $\boldsymbol{G}_{\mathrm{A}}$ are fixed, the cost function in (2.51) is quadratic in $\boldsymbol{\Lambda}$, and the optimal $\boldsymbol{\Lambda}$ is derived in closed-form by setting the derivative of the cost function to zero. When $\boldsymbol{Q}$ and $\boldsymbol{\Lambda}$ are fixed, the optimal $\boldsymbol{G}_{\mathrm{A}}$ is given by

$$\left[\boldsymbol{G}_{\mathrm{A}}^{\mathrm{H}}\right]_{i,j} = e^{\jmath \arg\left(\left[\boldsymbol{\Sigma}_{\mathrm{N}}^{-\frac{1}{2}}[\boldsymbol{U}_{\mathrm{NHN}}]_{:,N_r}\boldsymbol{\Lambda}\boldsymbol{Q}\right]_{i,j}\right)}. \tag{2.53}$$

When $\boldsymbol{G}_{\mathrm{A}}$ and $\boldsymbol{\Lambda}$ are fixed, the optimal unitary matrix $\boldsymbol{Q}$ is

$$\boldsymbol{Q} = \boldsymbol{U}\boldsymbol{V}^{\mathrm{H}}, \tag{2.54}$$

with $\boldsymbol{V}$ and $\boldsymbol{U}$ defined by the following SVD

$$\boldsymbol{G}_{\mathrm{A}}\boldsymbol{\Sigma}_{\mathrm{N}}^{-\frac{1}{2}}[\boldsymbol{U}_{\mathrm{NHN}}]_{:,N_r}\boldsymbol{\Lambda} = \boldsymbol{V}\boldsymbol{\Lambda}_3\boldsymbol{U}^{\mathrm{H}} \text{ with } \boldsymbol{\Lambda}_3 \searrow. \tag{2.55}$$

As proved in Appendix 1, the analog matrix $\boldsymbol{F}_{\mathrm{A}}$ of (2.46) is the solution of the following optimization problem

$$\max_{\boldsymbol{F}_{\mathrm{A}}} \boldsymbol{\lambda}\left(\left(\boldsymbol{F}_{\mathrm{A}}^{\mathrm{H}}\boldsymbol{F}_{\mathrm{A}}\right)^{-\frac{1}{2}}\boldsymbol{F}_{\mathrm{A}}^{\mathrm{H}}\boldsymbol{\Psi}_{\mathrm{H}}\boldsymbol{F}_{\mathrm{A}}\left(\boldsymbol{F}_{\mathrm{A}}^{\mathrm{H}}\boldsymbol{F}_{\mathrm{A}}\right)^{-\frac{1}{2}}\right). \tag{2.56}$$

The optimal $\boldsymbol{F}_{\mathrm{A}}$ can be computed by solving the following optimization problem

$$\begin{aligned} \min_{\boldsymbol{\Lambda},\boldsymbol{Q},\boldsymbol{F}_{\mathrm{A}}} \quad & \left\|[\boldsymbol{U}_{\boldsymbol{\Psi}_{\mathrm{H}}}]_{:,N_r}\boldsymbol{\Lambda}\boldsymbol{Q} - \boldsymbol{F}_{\mathrm{A}}\right\|_F^2 \\ \text{s.t.} \quad & \boldsymbol{Q}\boldsymbol{Q}^{\mathrm{H}} = \boldsymbol{I},\ \boldsymbol{F}_{\mathrm{A}} \in \mathcal{F}, \end{aligned} \tag{2.57}$$

where $\boldsymbol{U}_{\boldsymbol{\Psi}_{\mathrm{H}}}$ is defined in the EVD in (2.32). Similarly, an iterative optimization can be used to solve the optimization problem (2.57).

Since the objective function in (2.57) is quadratic in $\boldsymbol{\Lambda}$, the optimal $\boldsymbol{\Lambda}$ is readily derived in closed-form when $\boldsymbol{Q}$ and $\boldsymbol{F}_{\mathrm{A}}$ are fixed. When $\boldsymbol{Q}$ and $\boldsymbol{\Lambda}$ are fixed, the optimal $\boldsymbol{F}_{\mathrm{A}}$ is

$$[\boldsymbol{F}_{\mathrm{A}}]_{i,j} = e^{\jmath \arg\left(\left[[\boldsymbol{U}_{\boldsymbol{\Psi}_{\mathrm{H}}}]_{:,N_r}\boldsymbol{\Lambda}\boldsymbol{Q}\right]_{i,j}\right)}. \tag{2.58}$$

When $\boldsymbol{F}_{\mathrm{A}}$ and $\boldsymbol{\Lambda}$ are fixed, the optimal unitary matrix $\boldsymbol{Q}$ is $\boldsymbol{Q} = \boldsymbol{U}\boldsymbol{V}^{\mathrm{H}}$, with $\boldsymbol{V}$ and $\boldsymbol{U}$ defined in the following SVD

$$\boldsymbol{F}_{\mathrm{A}}^{\mathrm{H}}\left[\boldsymbol{U}_{\boldsymbol{\Psi}_{\mathrm{H}}}\right]_{:,N_t}\boldsymbol{\Lambda} = \boldsymbol{V}\boldsymbol{\Lambda}_4\boldsymbol{U}^{\mathrm{H}} \text{ with } \boldsymbol{\Lambda}_4 \searrow. \tag{2.59}$$

2.4.2 Sum MSE Minimization Based Training Optimization

Given the known analog matrices $\boldsymbol{G}_{\mathrm{A}}$ and $\boldsymbol{F}_{\mathrm{A}}$, the signal model in channel estimation is given by

$$\widetilde{\boldsymbol{Y}} = \boldsymbol{G}_{\mathrm{A}}\boldsymbol{H}\boldsymbol{F}_{\mathrm{A}}\boldsymbol{X} + \boldsymbol{G}_{\mathrm{A}}\boldsymbol{N}, \tag{2.60}$$

where $\boldsymbol{G}_{\mathrm{A}}$ and $\boldsymbol{F}_{\mathrm{A}}$ are both full rank squared matrices. Based on (2.60), we estimate the composite channel matrix $\boldsymbol{\mathcal{H}} = \boldsymbol{G}_{\mathrm{A}}\boldsymbol{H}\boldsymbol{F}_{\mathrm{A}}$, instead of $\boldsymbol{H}$, as

$$\widehat{\boldsymbol{\mathcal{H}}} = \widetilde{\boldsymbol{Y}}\boldsymbol{G}_{\mathrm{D}}, \tag{2.61}$$

with $\boldsymbol{G}_{\mathrm{D}}$ as the channel estimator. It is worth recapping that for this scheme to work, the analog matrices in channel estimation must be the same as in data transmission. The corresponding channel estimation MSE matrix is given by

$$\begin{aligned}&\boldsymbol{\Phi}_{\text{MSE}}(\boldsymbol{F}_{\text{A}},\boldsymbol{X},\boldsymbol{G}_{\text{A}},\boldsymbol{G}_{\text{D}})\\=&\mathbb{E}\{(\widetilde{\boldsymbol{Y}}\boldsymbol{G}_{\text{D}}-\boldsymbol{G}_{\text{A}}\boldsymbol{H}\boldsymbol{F}_{\text{A}})^{\text{H}}(\widetilde{\boldsymbol{Y}}\boldsymbol{G}_{\text{D}}-\boldsymbol{G}_{\text{A}}\boldsymbol{H}\boldsymbol{F}_{\text{A}})\}\\=&\boldsymbol{G}_{\text{A}}\boldsymbol{\Sigma}_{\text{H}}\boldsymbol{G}_{\text{A}}^{\text{H}}\text{Tr}\big(\boldsymbol{F}_{\text{A}}(\boldsymbol{X}\boldsymbol{G}_{\text{D}}-\boldsymbol{I})(\boldsymbol{X}\boldsymbol{G}_{\text{D}}-\boldsymbol{I})^{\text{H}}\boldsymbol{F}_{\text{A}}^{\text{H}}\boldsymbol{\Psi}_{\text{H}}\big)\\&+\boldsymbol{G}_{\text{A}}\boldsymbol{\Sigma}_{\text{N}}\boldsymbol{G}_{\text{A}}^{\text{H}}\text{Tr}\big(\boldsymbol{G}_{\text{D}}\boldsymbol{G}_{\text{D}}^{\text{H}}\boldsymbol{\Psi}_{\text{N}}\big),\end{aligned}\tag{2.62}$$

based on which the sum MSE can be expressed as

$$\begin{aligned}&\text{Tr}\big(\boldsymbol{\Phi}_{\text{MSE}}(\boldsymbol{F}_{\text{A}},\boldsymbol{X},\boldsymbol{G}_{\text{A}},\boldsymbol{G}_{\text{D}})\big)\\=&\text{Tr}\big(\boldsymbol{G}_{\text{A}}\boldsymbol{\Sigma}_{\text{H}}\boldsymbol{G}_{\text{A}}^{\text{H}}\big)\text{Tr}\big(\boldsymbol{F}_{\text{A}}(\boldsymbol{X}\boldsymbol{G}_{\text{D}}-\boldsymbol{I})(\boldsymbol{X}\boldsymbol{G}_{\text{D}}-\boldsymbol{I})^{\text{H}}\boldsymbol{F}_{\text{A}}^{\text{H}}\boldsymbol{\Psi}_{\text{H}}\big)\\&+\text{Tr}\big(\boldsymbol{G}_{\text{A}}\boldsymbol{\Sigma}_{\text{N}}\boldsymbol{G}_{\text{A}}^{\text{H}}\big)\text{Tr}\big(\boldsymbol{G}_{\text{D}}\boldsymbol{G}_{\text{D}}^{\text{H}}\boldsymbol{\Psi}_{\text{N}}\big).\end{aligned}\tag{2.63}$$

The term $\text{Tr}\big(\boldsymbol{\Phi}_{\text{MSE}}(\boldsymbol{F}_{\text{A}},\boldsymbol{X},\boldsymbol{G}_{\text{A}},\boldsymbol{G}_{\text{D}})\big)$ is a quadratic function for $\boldsymbol{G}_{\text{D}}$. Thus the optimal $\boldsymbol{G}_{\text{D}}$ satisfies the following equality

$$\begin{aligned}&\text{Tr}\big(\boldsymbol{G}_{\text{A}}\boldsymbol{\Sigma}_{\text{H}}\boldsymbol{G}_{\text{A}}^{\text{H}}\big)\boldsymbol{X}^{\text{H}}\boldsymbol{F}_{\text{A}}^{\text{H}}\boldsymbol{\Psi}_{\text{H}}\boldsymbol{F}_{\text{A}}\boldsymbol{X}\boldsymbol{G}_{\text{D}}+\text{Tr}\big(\boldsymbol{G}_{\text{A}}\boldsymbol{\Sigma}_{\text{N}}\boldsymbol{G}_{\text{A}}^{\text{H}}\big)\boldsymbol{\Psi}_{\text{N}}\boldsymbol{G}_{\text{D}}\\=&\text{Tr}\big(\boldsymbol{G}_{\text{A}}\boldsymbol{\Sigma}_{\text{H}}\boldsymbol{G}_{\text{A}}^{\text{H}}\big)\boldsymbol{X}^{\text{H}}\boldsymbol{F}_{\text{A}}^{\text{H}}\boldsymbol{\Psi}_{\text{H}}\boldsymbol{F}_{\text{A}},\end{aligned}\tag{2.64}$$

and it can be derived as

$$\begin{aligned}\boldsymbol{G}_{\text{D}}=&\Big(\text{Tr}\big(\boldsymbol{G}_{\text{A}}\boldsymbol{\Sigma}_{\text{H}}\boldsymbol{G}_{\text{A}}^{\text{H}}\big)\boldsymbol{X}^{\text{H}}\boldsymbol{F}_{\text{A}}^{\text{H}}\boldsymbol{\Psi}_{\text{H}}\boldsymbol{F}_{\text{A}}\boldsymbol{X}\\&+\text{Tr}\big(\boldsymbol{G}_{\text{A}}\boldsymbol{\Sigma}_{\text{N}}\boldsymbol{G}_{\text{A}}^{\text{H}}\big)\boldsymbol{\Psi}_{\text{N}}\Big)^{-1}\text{Tr}\big(\boldsymbol{G}_{\text{A}}\boldsymbol{\Sigma}_{\text{H}}\boldsymbol{G}_{\text{A}}^{\text{H}}\big)\boldsymbol{X}^{\text{H}}\boldsymbol{F}_{\text{A}}^{\text{H}}\boldsymbol{\Psi}_{\text{H}}\boldsymbol{F}_{\text{A}}.\end{aligned}\tag{2.65}$$

With the computed analog matrices $\boldsymbol{G}_{\text{A}}$ and $\boldsymbol{F}_{\text{A}}$, in the following, we will investigate the corresponding training optimizations for different performance metrics and different available channel statistics.

2.4.2.1 General Case of $\boldsymbol{\Psi}_{\text{N}} \not\propto \boldsymbol{I}$

When $\boldsymbol{G}_{\text{D}}$ is fixed, the sum MSE minimization based training optimization becomes

$$\begin{aligned}&\min_{\boldsymbol{X}}\ \text{Tr}\big(\boldsymbol{F}_{\text{A}}(\boldsymbol{X}\boldsymbol{G}_{\text{D}}-\boldsymbol{I})(\boldsymbol{X}\boldsymbol{G}_{\text{D}}-\boldsymbol{I})^{\text{H}}\boldsymbol{F}_{\text{A}}^{\text{H}}\boldsymbol{\Psi}_{\text{H}}\big)\\&\ \text{s.t.}\ \text{Tr}\big(\boldsymbol{F}_{\text{A}}\boldsymbol{X}\boldsymbol{X}^{\text{H}}\boldsymbol{F}_{\text{A}}^{\text{H}}\big)\le P.\end{aligned}\tag{2.66}$$

The corresponding Lagrange function is given by

$$\begin{aligned}\mathcal{L}(\boldsymbol{X},\mu)=&\text{Tr}\big(\boldsymbol{F}_{\text{A}}(\boldsymbol{X}\boldsymbol{G}_{\text{D}}-\boldsymbol{I})(\boldsymbol{X}\boldsymbol{G}_{\text{D}}-\boldsymbol{I})^{\text{H}}\boldsymbol{F}_{\text{A}}^{\text{H}}\boldsymbol{\Psi}_{\text{H}}\big)\\&+\mu\Big(\text{Tr}\big(\boldsymbol{F}_{\text{A}}\boldsymbol{X}\boldsymbol{X}^{\text{H}}\boldsymbol{F}_{\text{A}}^{\text{H}}\big)-P\Big),\end{aligned}\tag{2.67}$$

based on which the corresponding Karush-Kuhn-Tucker (KKT) conditions are given by [34]

$$\begin{cases} \boldsymbol{F}_{\mathrm{A}}^{\mathrm{H}}\boldsymbol{\Psi}_{\mathrm{H}}\boldsymbol{F}_{\mathrm{A}}\boldsymbol{X}\boldsymbol{G}_{\mathrm{D}}\boldsymbol{G}_{\mathrm{D}}^{\mathrm{H}} + \mu\boldsymbol{F}_{\mathrm{A}}^{\mathrm{H}}\boldsymbol{F}_{\mathrm{A}}\boldsymbol{X} = \boldsymbol{F}_{\mathrm{A}}^{\mathrm{H}}\boldsymbol{\Psi}_{\mathrm{H}}\boldsymbol{F}_{\mathrm{A}}\boldsymbol{G}_{\mathrm{D}}^{\mathrm{H}}, \\ \mu\Big(\mathrm{Tr}\big(\boldsymbol{F}_{\mathrm{A}}\boldsymbol{X}\boldsymbol{X}^{\mathrm{H}}\boldsymbol{F}_{\mathrm{A}}^{\mathrm{H}}\big) - P\Big) = 0, \\ \mu \geq 0, \ \mathrm{Tr}\big(\boldsymbol{F}_{\mathrm{A}}\boldsymbol{X}\boldsymbol{X}^{\mathrm{H}}\boldsymbol{F}_{\mathrm{A}}^{\mathrm{H}}\big) \leq P. \end{cases} \tag{2.68}$$

Based on the KKT conditions, the optimal training sequence $\boldsymbol{X}$ satisfies the following equality

$$\begin{aligned} \mathrm{vec}(\boldsymbol{X}) =& \Big(\big(\boldsymbol{G}_{\mathrm{D}}\boldsymbol{G}_{\mathrm{D}}^{\mathrm{H}}\big)^{\mathrm{T}} \otimes \boldsymbol{F}_{\mathrm{A}}^{\mathrm{H}}\boldsymbol{\Psi}_{\mathrm{H}}\boldsymbol{F}_{\mathrm{A}} + \mu\boldsymbol{I} \otimes \boldsymbol{F}_{\mathrm{A}}^{\mathrm{H}}\boldsymbol{F}_{\mathrm{A}}\Big)^{-1} \\ & \times \mathrm{vec}\big(\boldsymbol{F}_{\mathrm{A}}^{\mathrm{H}}\boldsymbol{\Psi}_{\mathrm{H}}\boldsymbol{F}_{\mathrm{A}}\boldsymbol{G}_{\mathrm{D}}^{\mathrm{H}}\big). \end{aligned} \tag{2.69}$$

As the optimization problem (2.66) is convex for $\boldsymbol{X}$, the KKT conditions (2.68) are the necessary and sufficient conditions for the optimal solutions. Then μ can be computed using one dimension search, e.g., bisection search, to guarantee that the KKT conditions hold [34].

2.4.2.2 Special Case of $\boldsymbol{\Psi}_{\mathrm{N}} = \alpha_{\mathrm{N}}\boldsymbol{I}$

Substituting $\boldsymbol{G}_{\mathrm{D}}$ of (2.65) into (2.63) yields

$$\mathrm{Tr}\big(\boldsymbol{\Phi}_{\mathrm{MSE}}(\boldsymbol{F}_{\mathrm{A}}, \boldsymbol{X}, \boldsymbol{G}_{\mathrm{A}}, \boldsymbol{G}_{\mathrm{D}})\big) = \mathrm{Tr}\left(\left(\frac{\big(\boldsymbol{F}_{\mathrm{A}}^{\mathrm{H}}\boldsymbol{\Psi}_{\mathrm{H}}\boldsymbol{F}_{\mathrm{A}}\big)^{-1}}{\mathrm{Tr}\big(\boldsymbol{G}_{\mathrm{A}}^{\mathrm{H}}\boldsymbol{\Sigma}_{\mathrm{H}}\boldsymbol{G}_{\mathrm{A}}\big)} + \frac{\boldsymbol{X}\boldsymbol{\Psi}_{\mathrm{N}}^{-1}\boldsymbol{X}^{\mathrm{H}}}{\mathrm{Tr}\big(\boldsymbol{G}_{\mathrm{A}}^{\mathrm{H}}\boldsymbol{\Sigma}_{\mathrm{N}}\boldsymbol{G}_{\mathrm{A}}\big)}\right)^{-1}\right), \tag{2.70}$$

where for composite channel estimation, both $\boldsymbol{G}_{\mathrm{A}}$ and $\boldsymbol{F}_{\mathrm{A}}$ are full rank squared matrices. Similar to the previous discussion, based on the MSE matrix formulation, the optimization for the hybrid MIMO system is given by

$$\begin{aligned} \min_{\boldsymbol{F}_{\mathrm{A}}, \boldsymbol{X}} \ & \mathrm{Tr}\left(\left(\frac{\big(\boldsymbol{F}_{\mathrm{A}}^{\mathrm{H}}\boldsymbol{\Psi}_{\mathrm{H}}\boldsymbol{F}_{\mathrm{A}}\big)^{-1}}{\mathrm{Tr}\big(\boldsymbol{G}_{\mathrm{A}}^{\mathrm{H}}\boldsymbol{\Sigma}_{\mathrm{H}}\boldsymbol{G}_{\mathrm{A}}\big)} + \frac{\boldsymbol{X}\boldsymbol{\Psi}_{\mathrm{N}}^{-1}\boldsymbol{X}^{\mathrm{H}}}{\mathrm{Tr}\big(\boldsymbol{G}_{\mathrm{A}}^{\mathrm{H}}\boldsymbol{\Sigma}_{\mathrm{N}}\boldsymbol{G}_{\mathrm{A}}\big)}\right)^{-1}\right) \\ \text{s.t.} \ & \mathrm{Tr}\big(\boldsymbol{F}_{\mathrm{A}}\boldsymbol{X}\boldsymbol{X}^{\mathrm{H}}\boldsymbol{F}_{\mathrm{A}}^{\mathrm{H}}\big) \leq P. \end{aligned} \tag{2.71}$$

Similarly, by defining the following auxiliary variable $\widetilde{\boldsymbol{X}}$

$$\widetilde{\boldsymbol{X}} = \big(\boldsymbol{F}_{\mathrm{A}}^{\mathrm{H}}\boldsymbol{F}_{\mathrm{A}}\big)^{\frac{1}{2}}\boldsymbol{X}, \tag{2.72}$$

the optimization problem (2.71) is rewritten as

$$\begin{aligned}
&\min_{\boldsymbol{F}_{\mathrm{A}},\widetilde{\boldsymbol{X}}} \operatorname{Tr}\left(\left(\left(\boldsymbol{F}_{\mathrm{A}}^{\mathrm{H}}\boldsymbol{\Psi}_{\mathrm{H}}\boldsymbol{F}_{\mathrm{A}}\right)^{-1}+\alpha_2\left(\boldsymbol{F}_{\mathrm{A}}^{\mathrm{H}}\boldsymbol{F}_{\mathrm{A}}\right)^{-\frac{1}{2}}\widetilde{\boldsymbol{X}}\widetilde{\boldsymbol{X}}^{\mathrm{H}}\left(\boldsymbol{F}_{\mathrm{A}}^{\mathrm{H}}\boldsymbol{F}_{\mathrm{A}}\right)^{-\frac{1}{2}}\right)^{-1}\right)\\
&\text{s.t. } \operatorname{Tr}\left(\widetilde{\boldsymbol{X}}\widetilde{\boldsymbol{X}}^{\mathrm{H}}\right)\le P,
\end{aligned} \tag{2.73}$$

where

$$\alpha_2 = \operatorname{Tr}(\boldsymbol{G}_{\mathrm{A}}^{\mathrm{H}}\boldsymbol{\Sigma}_{\mathrm{H}}\boldsymbol{G}_{\mathrm{A}})/[\alpha_{\mathrm{N}}\operatorname{Tr}(\boldsymbol{G}_{\mathrm{A}}^{\mathrm{H}}\boldsymbol{\Sigma}_{\mathrm{N}}\boldsymbol{G}_{\mathrm{A}})]. \tag{2.74}$$

To transfer this optimization problem into a convex one that can be efficiently solved, we introduce a new variable $\boldsymbol{Q}$ as

$$\boldsymbol{Q} = \widetilde{\boldsymbol{X}}\widetilde{\boldsymbol{X}}^{\mathrm{H}}. \tag{2.75}$$

Since $\widetilde{\boldsymbol{X}}$ is a fat matrix, to guarantee that the channel matrix can be estimated, $\boldsymbol{Q}$ is a full rank matrix. Thus, in the optimization, we do not need to consider the rank constraint on $\boldsymbol{Q}$. With the definition of $\boldsymbol{Q}$, (2.73) is equivalent to [34]

$$\begin{aligned}
&\min_{\boldsymbol{M},\boldsymbol{Q}} \operatorname{Tr}(\boldsymbol{M})\\
&\text{s.t. } \boldsymbol{M}\succeq\left(\left(\boldsymbol{F}_{\mathrm{A}}^{\mathrm{H}}\boldsymbol{\Psi}_{\mathrm{H}}\boldsymbol{F}_{\mathrm{A}}\right)^{-1}+\alpha_2\left(\boldsymbol{F}_{\mathrm{A}}^{\mathrm{H}}\boldsymbol{F}_{\mathrm{A}}\right)^{-\frac{1}{2}}\boldsymbol{Q}\left(\boldsymbol{F}_{\mathrm{A}}^{\mathrm{H}}\boldsymbol{F}_{\mathrm{A}}\right)^{-\frac{1}{2}}\right)^{-1}\\
&\quad\ \boldsymbol{M}\succeq\boldsymbol{0},\ \boldsymbol{Q}\succeq\boldsymbol{0},\ \operatorname{Tr}(\boldsymbol{Q})\le P,
\end{aligned} \tag{2.76}$$

where $\boldsymbol{M}$ is a positive semidefinite matrix. Then using Schur complement, the first constraint can be replaced by the following linear matrix inequality equivalently,

$$\begin{bmatrix} \boldsymbol{M} & \boldsymbol{I}\\ \boldsymbol{I} & \left(\boldsymbol{F}_{\mathrm{A}}^{\mathrm{H}}\boldsymbol{\Psi}_{\mathrm{H}}\boldsymbol{F}_{\mathrm{A}}\right)^{-1}+\alpha_2\left(\boldsymbol{F}_{\mathrm{A}}^{\mathrm{H}}\boldsymbol{F}_{\mathrm{A}}\right)^{-\frac{1}{2}}\boldsymbol{Q}\left(\boldsymbol{F}_{\mathrm{A}}^{\mathrm{H}}\boldsymbol{F}_{\mathrm{A}}\right)^{-\frac{1}{2}}\end{bmatrix}\succeq\boldsymbol{0}. \tag{2.77}$$

As a result, the optimization (2.73) can be transferred into the standard semidefinite programming (SDP) problem given as

$$\begin{aligned}
&\min_{\boldsymbol{M},\boldsymbol{Q}} \operatorname{Tr}(\boldsymbol{M})\\
&\text{s.t. } \begin{bmatrix} \boldsymbol{M} & \boldsymbol{I}\\ \boldsymbol{I} & \left(\boldsymbol{F}_{\mathrm{A}}^{\mathrm{H}}\boldsymbol{\Psi}_{\mathrm{H}}\boldsymbol{F}_{\mathrm{A}}\right)^{-1}+\alpha_2\left(\boldsymbol{F}_{\mathrm{A}}^{\mathrm{H}}\boldsymbol{F}_{\mathrm{A}}\right)^{-\frac{1}{2}}\boldsymbol{Q}\left(\boldsymbol{F}_{\mathrm{A}}^{\mathrm{H}}\boldsymbol{F}_{\mathrm{A}}\right)^{-\frac{1}{2}}\end{bmatrix}\succeq\boldsymbol{0},\ \boldsymbol{Q}\succeq\boldsymbol{0},
\end{aligned} \tag{2.78}$$

which can be solved efficiently using, for example, the CVX software toolbox, a Matlab software for disciplined convex programming.

2.4.3 MI Maximization Based Training Optimization

For the composite channel estimation, the training optimization problem based on MI maximization can be formulated in the following form,

$$\begin{aligned} \max_{\boldsymbol{X},\boldsymbol{G}_{\mathrm{D}}} \quad & -\log|\boldsymbol{\Phi}_{\mathrm{MSE}}(\boldsymbol{F}_{\mathrm{A}},\boldsymbol{X},\boldsymbol{G}_{\mathrm{A}},\boldsymbol{G}_{\mathrm{D}})| \\ \text{s.t.} \quad & \mathrm{Tr}\big(\boldsymbol{F}_{\mathrm{A}}\boldsymbol{X}\boldsymbol{X}^{\mathrm{H}}\boldsymbol{F}_{\mathrm{A}}^{\mathrm{H}}\big) \le P. \end{aligned} \tag{2.79}$$

Substituting (2.65) into (2.62) and using the resultant (2.62) in (2.79), we have the following equivalent optimization problem

$$\begin{aligned} \max_{\boldsymbol{X}} \; & log\left|\big(\boldsymbol{F}_{\mathrm{A}}^{\mathrm{H}}\boldsymbol{\Psi}_{\mathrm{H}}\boldsymbol{F}_{\mathrm{A}}\big)^{-1}+\alpha_2\boldsymbol{X}\boldsymbol{X}^{\mathrm{H}}\right| \\ \text{s.t.} \; & \mathrm{Tr}\big(\boldsymbol{F}_{\mathrm{A}}\boldsymbol{X}\boldsymbol{X}^{\mathrm{H}}\boldsymbol{F}_{\mathrm{A}}^{\mathrm{H}}\big) \le P. \end{aligned} \tag{2.80}$$

In contrast to the sum MSE minimization, the MI maximization in nature minimizes the geometric mean of the diagonal elements of the channel estimation MSE matrix.

Similarly after defining $\widetilde{\boldsymbol{X}}=\big(\boldsymbol{F}_{\mathrm{A}}^{\mathrm{H}}\boldsymbol{F}_{\mathrm{A}}\big)^{\frac{1}{2}}\boldsymbol{X}$, we have

$$\begin{aligned} \max_{\widetilde{\boldsymbol{X}}} \; & log\left|\big(\boldsymbol{F}_{\mathrm{A}}^{\mathrm{H}}\boldsymbol{\Psi}_{\mathrm{H}}\boldsymbol{F}_{\mathrm{A}}\big)^{-1}+\alpha_2\big(\boldsymbol{F}_{\mathrm{A}}^{\mathrm{H}}\boldsymbol{F}_{\mathrm{A}}\big)^{-\frac{1}{2}}\widetilde{\boldsymbol{X}}\widetilde{\boldsymbol{X}}^{\mathrm{H}}\big(\boldsymbol{F}_{\mathrm{A}}^{\mathrm{H}}\boldsymbol{F}_{\mathrm{A}}\big)^{-\frac{1}{2}}\right| \\ \text{s.t.} \; & \mathrm{Tr}\big(\widetilde{\boldsymbol{X}}\widetilde{\boldsymbol{X}}^{\mathrm{H}}\big) \le P, \end{aligned} \tag{2.81}$$

which is equivalent to the following optimization problem

$$\begin{aligned} \max_{\widetilde{\boldsymbol{X}}} \; & log\left|\big(\boldsymbol{F}_{\mathrm{A}}^{\mathrm{H}}\boldsymbol{F}_{\mathrm{A}}\big)^{\frac{1}{2}}\big(\boldsymbol{F}_{\mathrm{A}}^{\mathrm{H}}\boldsymbol{\Psi}_{\mathrm{H}}\boldsymbol{F}_{\mathrm{A}}\big)^{-1}\big(\boldsymbol{F}_{\mathrm{A}}^{\mathrm{H}}\boldsymbol{F}_{\mathrm{A}}\big)^{\frac{1}{2}}+\alpha_2\widetilde{\boldsymbol{X}}\widetilde{\boldsymbol{X}}^{\mathrm{H}}\right| \\ \text{s.t.} \; & \mathrm{Tr}\big(\widetilde{\boldsymbol{X}}\widetilde{\boldsymbol{X}}^{\mathrm{H}}\big) \le P. \end{aligned} \tag{2.82}$$

It is obvious that the optimization (2.82) has closed-form optimal solutions. Specifically, based on the inequality (2.39), the optimal $\widetilde{\boldsymbol{X}}$ satisfies

$$\widetilde{\boldsymbol{X}}=\boldsymbol{Q}_{\boldsymbol{F}}\boldsymbol{\Lambda}_{\widetilde{\boldsymbol{X}}}\boldsymbol{U}_{\mathrm{Arb}}^{\mathrm{H}}, \tag{2.83}$$

where $\boldsymbol{\Lambda}_{\widetilde{\boldsymbol{X}}}$ is a diagonal matrix, and $\boldsymbol{U}_{\mathrm{Arb}}$ is an arbitrary unitary matrix with proper dimension, while the unitary matrix $\boldsymbol{Q}_{\boldsymbol{F}}$ is defined based on the following EVD

$$\big(\boldsymbol{F}_{\mathrm{A}}^{\mathrm{H}}\boldsymbol{F}_{\mathrm{A}}\big)^{\frac{1}{2}}\big(\boldsymbol{F}_{\mathrm{A}}^{\mathrm{H}}\boldsymbol{\Psi}_{\mathrm{H}}\boldsymbol{F}_{\mathrm{A}}\big)^{-1}\big(\boldsymbol{F}_{\mathrm{A}}^{\mathrm{H}}\boldsymbol{F}_{\mathrm{A}}\big)^{\frac{1}{2}}=\boldsymbol{Q}_{\boldsymbol{F}}\boldsymbol{\Lambda}_{\boldsymbol{F}\boldsymbol{\Psi}\boldsymbol{F}}\boldsymbol{Q}_{\boldsymbol{F}}^{\mathrm{H}}\text{with } \boldsymbol{\Lambda}_{\boldsymbol{F}\boldsymbol{\Psi}\boldsymbol{F}}\searrow. \tag{2.84}$$

Based on the optimal structure (2.83), the original optimization problem is simplified into

$$\begin{aligned} \max_{\{x_i^2\}} \ & \sum_i \log(\lambda_i + \alpha_2 x_i^2) \\ \text{s.t.} \ & \sum_i x_i^2 \le P, \end{aligned} \tag{2.85}$$

whose optimal solution is the water-filling solution [34]

$$x_i^2 = (1/\mu - \lambda_i/\alpha_2)^+ , \tag{2.86}$$

where $\lambda_i = [\mathbf{\Lambda}_{F\Psi F}]_{i,i}$.

2.4.4 Complexity Analysis

Firstly, it may be readily observed that both benchmarks have the same complexity order as the proposed LS estimator of Sect. 2.3.1, since the same DFT-based training sequence is adopted, which is dominated by the matrix inversion operation having a complexity order of $\mathcal{O}(N_t^3 + N_r^3)$, where N_t and N_r denote the numbers of antennas at the transmitter and the receiver, respectively. However, in practice, both benchmarks have more complex signal processing operations than the LS estimator, since they need either partial or complete channel statistics for channel estimation. Additionally, considering that both the matrix inversion and EVD operations are involved in deriving the optimal training sequence for the proposed MMSE estimator of Sect. 2.3.2, the complexity order is also $\mathcal{O}(N_t^3 + N_r^3)$, whilst achieving a better MSE performance than the proposed LS estimator. In Sect. 2.4, where the composite channel estimation having drastically reduced dimensions of $N_{RF} \times N_{RF}$ ($N_{RF} \ll \min(N_t, N_r)$) is studied, the complexity order of the proposed iterative MMSE (MMSE-ITE) estimator detailed in Sect. 2.4.2.1 is $\mathcal{O}[I(N_{RF}L)^3]$, where N_{RF} and L denote the number of RF chains and the training sequence length, respectively, while I represents the number of iterations required for solving the problem (2.66). Moreover, the proposed MMSE estimator of Sect. 2.4.2.2 is obtained by solving the SDP problem (2.78), whose complexity order is $\mathcal{O}(N_t^{3.5})\log(1/\epsilon)$ with ϵ being the precision factor. Note that the above complexity analysis of all the proposed channel estimators is also applicable to the associated MI maximization counterparts.

2.5 Numerical Results and Analysis

In this section, without loss of generality, the widely used Raleigh fading channel model is adopted. We numerically evaluate the normalized MSE with respect to the effective channel statistics $\mathrm{Tr}(\boldsymbol{R}_{\mathrm{H}})$ and the MI performance of the LS estimator in Sect. 2.3.1, the MMSE, i.e., LMMSE, estimator in Sect. 2.3.2

and the MMSE, i.e., LMMSE, estimator in Sect. 2.4, respectively, under a wide range of SNRs, antenna setups, the number of RF chains, and different channel correlations. Note that the channel covariance matrix $\boldsymbol{R}_{\mathrm{H}}$ is different for different channel estimators. Specifically, for the LS estimator of Sect. 2.3.1, $\boldsymbol{R}_{\mathrm{H}}=\mathbb{E}\{\boldsymbol{H}\boldsymbol{H}^{\mathrm{H}}\}=\mathrm{Tr}(\boldsymbol{\Psi}_{\mathrm{H}})\boldsymbol{\Sigma}_{\mathrm{H}}$. For the MMSE estimator of Sect. 2.3.2, $\boldsymbol{R}_{\mathrm{H}}=\mathbb{E}\{\boldsymbol{G}_{\mathrm{A}}\boldsymbol{H}\boldsymbol{H}^{\mathrm{H}}\boldsymbol{G}_{\mathrm{A}}^{\mathrm{H}}\}=\mathrm{Tr}(\boldsymbol{\Psi}_{\mathrm{H}})\boldsymbol{G}_{\mathrm{A}}\boldsymbol{\Sigma}_{\mathrm{H}}\boldsymbol{G}_{A}^{\mathrm{H}}$, while for the MMSE estimator of Sect. 2.4, $\boldsymbol{R}_{\mathrm{H}}=\mathbb{E}\{\boldsymbol{G}_{\mathrm{A}}\boldsymbol{H}\boldsymbol{F}_{\mathrm{A}}\boldsymbol{F}_{\mathrm{A}}^{\mathrm{H}}\boldsymbol{H}^{\mathrm{H}}\boldsymbol{G}_{\mathrm{A}}^{\mathrm{H}}\}=\mathrm{Tr}(\boldsymbol{F}_{\mathrm{A}}^{\mathrm{H}}\boldsymbol{\Psi}_{\mathrm{H}}\boldsymbol{F}_{\mathrm{A}})\boldsymbol{G}_{\mathrm{A}}\boldsymbol{\Sigma}_{\mathrm{H}}\boldsymbol{G}_{\mathrm{A}}^{\mathrm{H}}$. Moreover, the scaled LS (SLS) estimator [14] and the MMSE estimator with orthogonal probing based on DFT (MMSE-DFT) [13, 14] are adopted as two comparisons. Note that except for the LS estimator, all the other estimators require the second-order statistics of the channel and noise at the receiver. Furthermore, for the optimal training design, the eigenspace of the channel and noise second-order statistics is assumed to be known at the transmitter. Unfortunately, these two requirements impose large computational cost and high feedback overhead.

Unless otherwise stated, the numbers of transmit and receive antennas are set to $N_t=32$, and $N_r=16$, respectively, and the number of RF chains at both the transmitter and receiver is $N_{RF}=4$. Moreover, the training length is set to $L=N_t=32$ for both the LS estimator of Sect. 2.3.1 and the MMSE estimator of Sect. 2.3.2, and $L=N_{RF}=4$ for the MMSE estimator of Sect. 2.4, respectively. The received noise power is assumed to be unity and hence the training SNR is defined as SNR $=P$. We also utilize the exponential model to construct the correlation matrices of the channel and noise as follows

$$\begin{cases} [\boldsymbol{\Sigma}_{\mathrm{H}}]_{n_1,m_1} = |\zeta_t|^{|n_1-m_1|}, & [\boldsymbol{\Sigma}_{\mathrm{N}}]_{n_1,m_1} = |\zeta_r|^{|n_1-m_1|}, \\ [\boldsymbol{\Psi}_{\mathrm{H}}]_{n_2,m_2} = |\epsilon_t|^{|n_2-m_2|}, & [\boldsymbol{\Psi}_{\mathrm{N}}]_{n_2,m_2} = |\epsilon_r|^{|n_2-m_2|}, \\ |\zeta_t| \le 1, \quad |\zeta_r| \le 1, \quad |\epsilon_t| \le 1, \quad |\epsilon_r| \le 1, \end{cases} \tag{2.87}$$

where ζ_t, ζ_r, ϵ_t and ϵ_r are the correlation coefficients. Unless otherwise stated, we set $\zeta_t = \epsilon_t = 0.5$ and $\zeta_r = \epsilon_r = 0.4$.

2.5.1 Normalized MSE Performance

Figure 2.2 shows the normalized MSE performance of the LS and SLS estimators as functions of the SNR, where three different antenna setups of $N_t \times N_r = 16 \times 16$, 32×16 and 64×16 are considered. As expected, the normalized MSE performance of both estimators improves as the SNR increases. In addition, the SLS estimator with the aid of the channel and noise second-order statistics naturally performs better than the LS estimator, especially in low-SNR regions. However, this performance gain is realized at the expense of significantly higher computational cost.

Figure 2.3 illustrates the normalized MSE performance of the LS, SLS, and MMSE of Sect. 2.3.2 estimators as functions of the SNR, under two different antenna setups of $N_t \times N_r = 16 \times 16$ and 32×16. Observe that for both these two antenna setups,

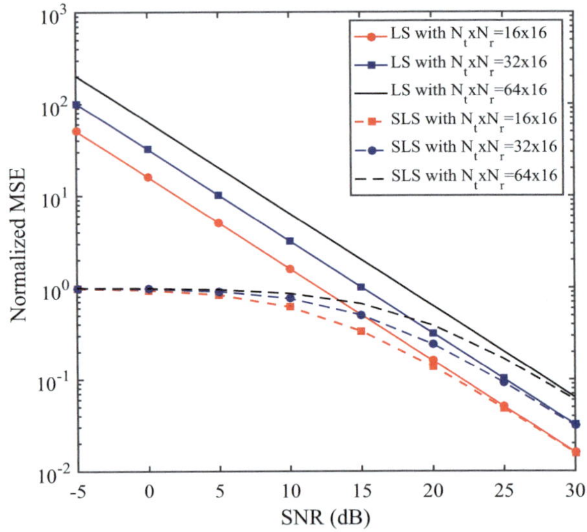

Fig. 2.2 Normalized MSE performance of the proposed LS and benchmark SLS estimators as functions of SNR under different antenna setups of $N_t \times N_r = 16 \times 16$, 32×16 and 64×16

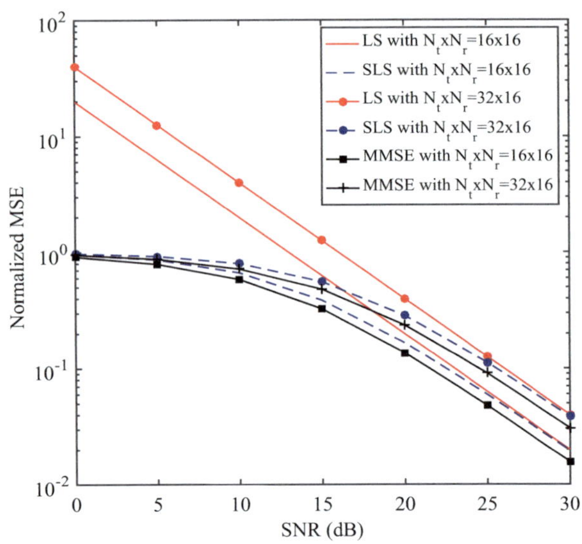

Fig. 2.3 Normalized MSE performance of the proposed LS, and MMSE of Sect. 2.3.2 as well as benchmark SLS estimators as functions of SNR under different antenna setups of $N_t \times N_r = 16 \times 16$ and 32×16

the proposed MMSE estimator achieves better MSE performance than the benchmark SLS, since it directly minimizes the channel estimation error by utilizing more channel and noise second-order statistics compared to the SLS estimator. The performance of the proposed LS estimator is the worst, as it does not utilize any seconder statistics of the channel and noise and imposes the lowest complexity.

Figure 2.4 depicts the normalized MSEs as functions of the SNR achieved by the MMSE estimator with optimal training of Sect. 2.3.2, and the benchmark MMSE-DFT with orthogonal training, under three different antenna setups together with two different channel correlations of $\epsilon_t = 0.3$ and $\epsilon_t = 0.8$, respectively. It can be seen from Fig. 2.4a that with a weakly correlated channel of $\epsilon_t = 0.3$, the performance of the MMSE-DFT is almost identical to that of our optimal MMSE estimator. This is because the orthogonal training is nearly optimal when the channel is asymptotically uncorrelated. However, observe from Fig. 2.4b that with $\epsilon_t = 0.8$, the performance gap between the MMSE-DFT benchmark and our optimal MMSE estimator is clearly visible, particularly at low SNR region.

Similarly to Figs. 2.4 and 2.5 compares the normalized MSEs of the proposed MMSE estimator of Sect. 2.4.2.1 and the MMSE-DFT benchmark by varying SNR, where two antenna setups and two cases of channel correlation are considered. Different from Sect. 2.3.2, since the dimension of the estimated composite channel $\boldsymbol{G}_{\mathrm{A}}\boldsymbol{H}\boldsymbol{F}_{\mathrm{A}}$ in Sect. 2.4 does not depend on the number of transmit antennas, increasing N_t actually provides the gain in the received SNR. Consequently, with the increase of N_t, it can be seen from Fig. 2.5 that the MSEs of the two estimators become better. In addition, the performance gap between the two estimators for the strong channel correlation is larger than that for the weak counterpart.

Figure 2.6 further depicts the normalized MSEs achieved by the MMSE-ITE estimator of Sect. 2.4.2.1, MMSE estimator of Sect. 2.4.2.2 and benchmark MMSE-DFT under the same conditions of Fig. 2.6 but with $\epsilon_r = 0$. In particular, we find that the proposed MMSE-ITE estimator of Sect. 2.4.2.1 can achieve nearly the same performance as the optimal MMSE estimator of Sect. 2.4.2.2 for both weak and strong

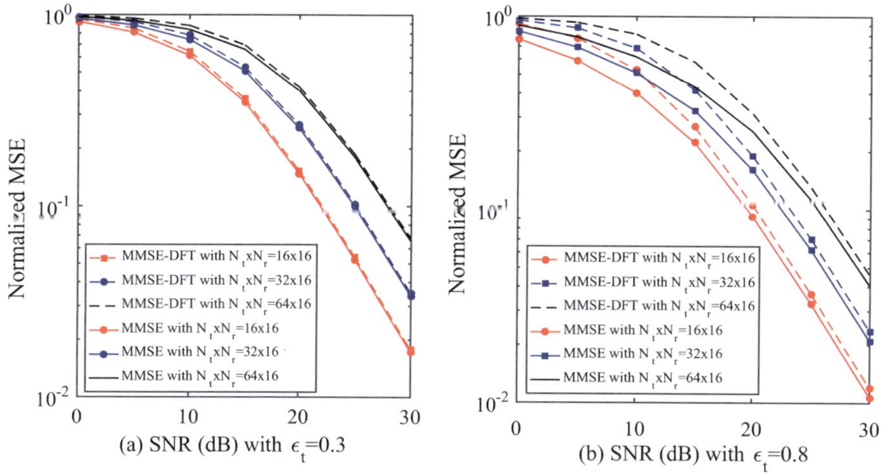

Fig. 2.4 Normalized MSE performance of proposed MMSE of Sect. 2.3.2 and benchmark MMSE-DFT estimators as functions of SNR under different antenna setups of $N_t \times N_r = 16 \times 16$, 32×16 and 64×16 with **a** weak channel correlation $\epsilon_t = 0.3$, and **b** strong channel correlation $\epsilon_t = 0.8$

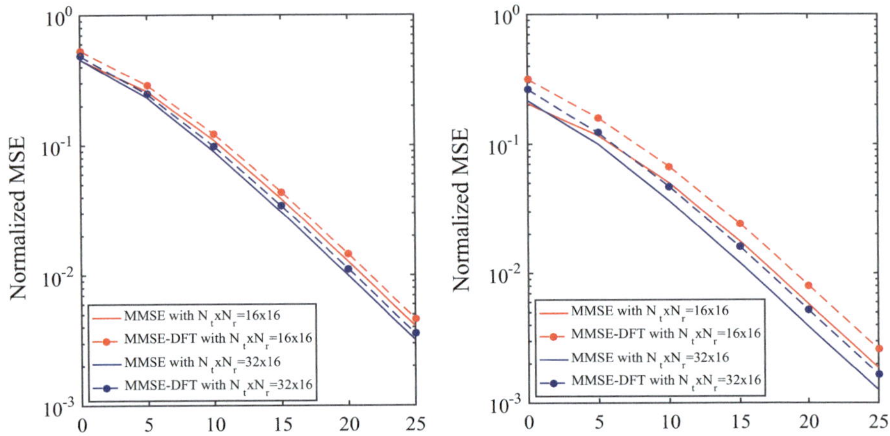

(a) SNR (dB) with ϵ_t=0.3 (b) SNR (dB) with ϵ_t=0.8

Fig. 2.5 Normalized MSE performance of the proposed MMSE (of Sect. 2.4.2.1) and benchmark MMSE-DFT estimators as functions of SNR under different antenna setups of $N_t \times N_r = 16 \times 16$ and 32×16 with (a) weak channel correlation $\epsilon_t = 0.3$, and (b) strong channel correlation $\epsilon_t = 0.8$

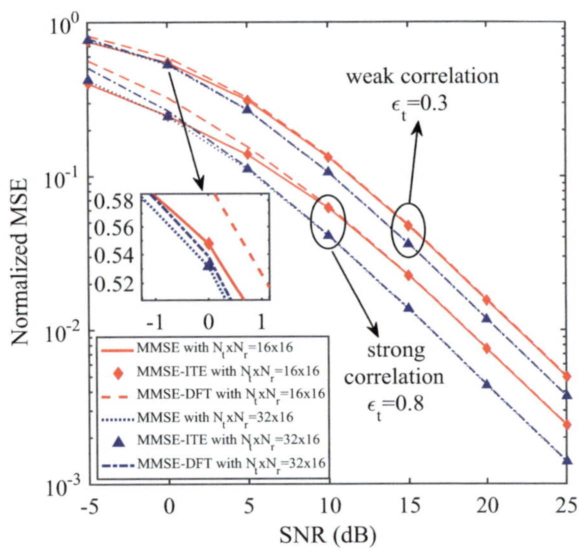

Fig. 2.6 Normalized MSE performance of the proposed MMSE-ITE estimator of Sect. 2.4.2.1 and MMSE estimator of Sect. 2.4.2.2 as well as benchmark MMSE-DFT as functions of SNR under different antenna setups of $N_t \times N_r = 16 \times 16$ and 32×16 as well as different channel correlations of $\epsilon_t = 0.3$, $\epsilon_t = 0.8$ and $\epsilon_r = 0$

channel correlations. As expected, the two proposed MMSE estimators outperform the MMSE-DFT benchmark, especially for the case of $N_t = 16$ and low SNR.

It is readily observed from Sect. 2.3.2 that the number of RF chains N_{RF} actually has no influence on the proposed MMSE estimator since the analog matrix $\boldsymbol{G}_A$ is assumed to be column full rank and thus the corresponding sum-MSE minimization in (2.23) is irrelevant to N_{RF}. However, we readily find that by setting $\boldsymbol{G}_{D,R} = \boldsymbol{I}_{N_{RF}}$,

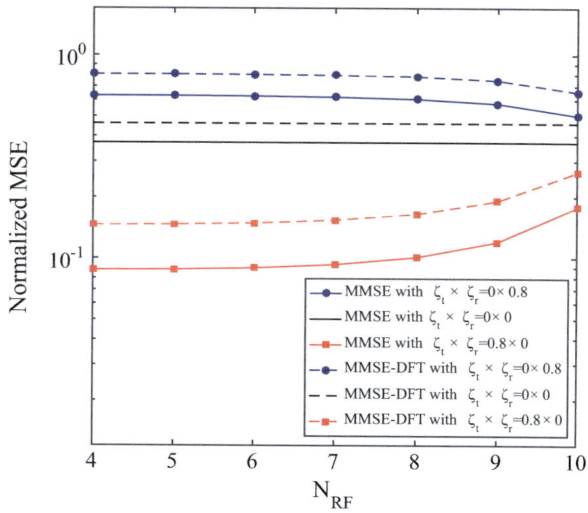

Fig. 2.7 Normalized MSE performance of the proposed MMSE (of Sect. 2.3.2) and benchmark MMSE-DFT estimators as functions of the number of RF chains N_{RF} under different channel and noise correlations of $\zeta_t \times \zeta_r = 0 \times 0$, 0×0.8 and 0.8×0, given SNR = 15 dB

the proposed MMSE estimator of Sect. 2.3.2 can also be applied to estimate the effective channel $\boldsymbol{G}_A \boldsymbol{H}$ with much reduced channel dimension $N_{RF} \times N_t$. In this context, we can reexpress α_1 in (2.21) as $\alpha_1 = \mathrm{Tr}\big(\boldsymbol{G}_{\mathrm{A}} \boldsymbol{\Sigma}_{\mathrm{H}} \boldsymbol{G}_{\mathrm{A}}^{\mathrm{H}}\big) / \mathrm{Tr}\big(\boldsymbol{G}_{\mathrm{A}} \boldsymbol{\Sigma}_{\mathrm{N}} \boldsymbol{G}_{\mathrm{A}}^{\mathrm{H}}\big)$. Based on this setting, in Fig. 2.7, the normalized MSEs of the optimal training-based MMSE estimator derived in Sect. 2.3.2 and the benchmark MMSE-DFT are compared by varying N_{RF}, given SNR $= 15$ dB and different channel and noise correlations. It is easily inferred from (2.21) that the influence of N_{RF} on the MSE is mainly dominated by the parameter α_1. We consider three typical cases of α_1 as: 1) $\zeta_t = \zeta_r = 0$, 2) $\zeta_t = 0$ and $\zeta_r = 0.8$, and 3) $\zeta_t = 0.8$ and $\zeta_r = 0$. In the first case, N_{RF} has no influence on the achievable MSEs of the two estimators, due to the fixed $\alpha_1 = 1$ when varying N_{RF}. In the second case, the performances of the two estimators become better for large N_{RF}, since it can be inferred that the value of α_1 increases for a large N_{RF}. Conversely, in the third case, the MSEs of the two estimators become poorer for large N_{RF}, due to the decreased value of α_1 corresponding to a large N_{RF}. Naturally, the proposed MMSE estimator always outperforms the MMSE-DFT benchmark in terms of MSE performance.

Figure 2.8 compares the normalized MSEs of the MMSE estimator of Sect. 2.4.2.1 and benchmark MMSE-DFT by varying N_{RF}, under two antenna setups and given SNR = 15 dB. Observe from Fig. 2.8 that as N_{RF} increases, both estimators have worse performance, due to the increased dimension of the composite channel $\boldsymbol{G}_{\mathrm{A}} \boldsymbol{H} \boldsymbol{F}_{\mathrm{A}}$. In particular, the MMSE-DFT estimator achieves a comparable performance to the MMSE estimator at low N_{RF}, since in this case the optimal training with the decreased design freedom is asymptotically orthogonal. Like Fig. 2.6, increasing N_t also leads to better performance.

Figure 2.9 shows the MI performance of the MMSE estimator of Sect. 2.3.2 and benchmark MMSE-DFT versus the SNR, under three different antenna setups and two cases of channel correlation. Firstly, we observe that increasing N_t leads to

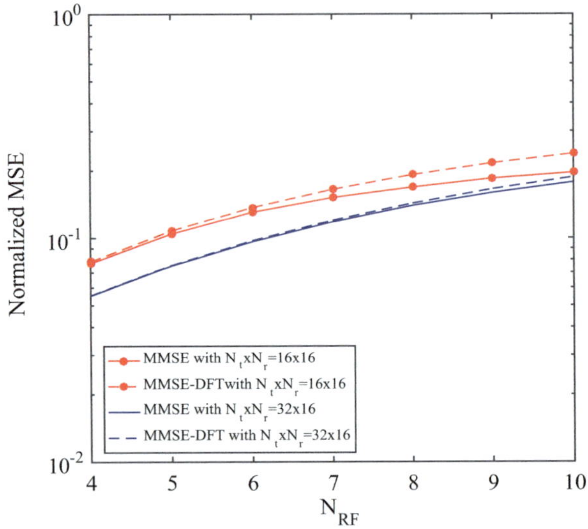

Fig. 2.8 Normalized MSEs of the MMSE estimator of Sect. 2.4.2.1 and benchmark MMSE-DFT versus the number of RF chains under different antenna setups of $N_t \times N_r = 16 \times 16$ and 32×16 with SNR = 15 dB

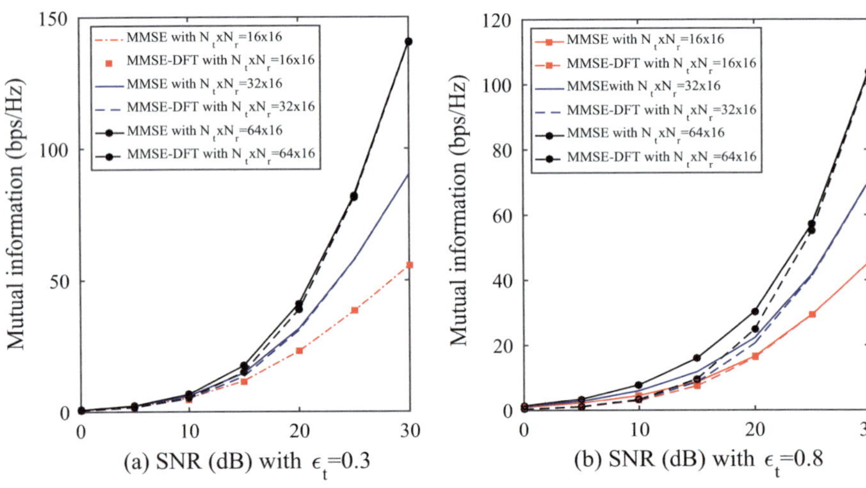

Fig. 2.9 Achievable MI performance of the MMSE estimator of Sect. 2.3.2 and benchmark MMSE-DFT versus the SNR, under different antenna setups of $N_t \times N_r = 16 \times 16$, 32×16 and 64×16 with (a) weak channel correlation of $\epsilon_t = 0.3$, and (b) strong channel correlation of $\epsilon_t = 0.8$

better MI performance for the two estimators. Moreover, the MI performance of the two estimators is almost identical at a low N_t value and/or high SNR. Comparing Figs. 2.9a, b, the MI performance gap between the two estimators for the weak channel correlation is clearly narrower than that for the strong counterpart, especially for large N_t value.

We also extend this simulation to the composite channel estimation and compare the MI performance of the MMSE estimator proposed in Sect. 2.4.2.1 with the benchmark MMSE-DFT in Fig. 2.10. It is readily seen that the conclusions obtained in Fig. 2.9 are still applicable to Fig. 2.10. Furthermore, considering the much-reduced channel dimensions to be estimated in Sect. 2.4, the achievable MI values of the two estimators are evidently lower than those of Fig. 2.9.

In summary, the proposed LS estimator requiring no channel statistics performs close to the proposed MMSE estimator of Sect. 2.3.2 relying on the second-order statistics of channel and noise for both weak and strong channel correlations. This is particularly so at high SNR. By contrast in the low-SNR regime, it can be concluded that for the case of weak channel correlation, all the proposed MMSE estimators of Sects. 2.3.2, 2.4.2.1, and 2.4.2.2 exhibit a similar performance to the corresponding MMSE-DFT estimator. In this case, it would be preferable to apply the low-complexity MMSE-DFT estimator. In the case of strong channel correlation, all the proposed MMSE estimators notably outperform the MMSE-DFT estimator and thus are preferred. Conversely, the proposed MMSE estimator of Sect. 2.4.2.1 used for estimating the composite channel $\boldsymbol{G}_A \boldsymbol{H} \boldsymbol{F}_A$ is recommended in case of a large number of RF chains. Similarly, all the above conclusions for our proposed MMSE estimators still hold for the MI criterion.

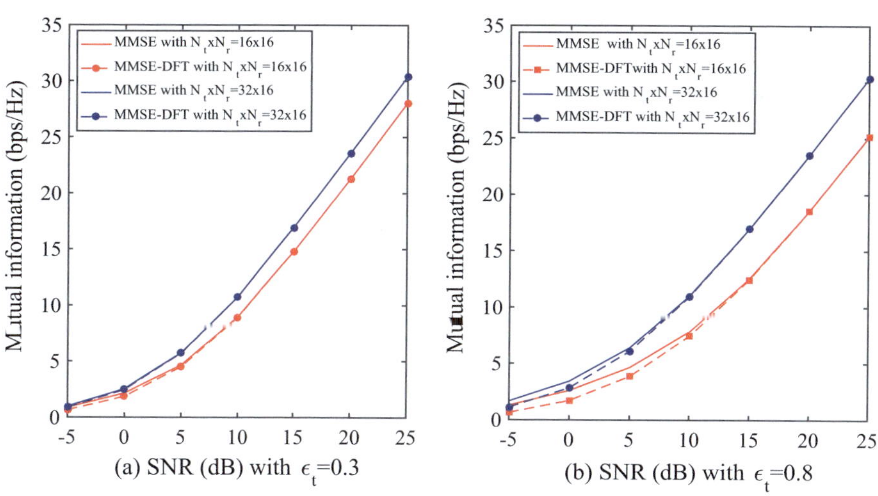

Fig. 2.10 Achievable MI performance of the MMSE estimator of Sect. 2.4.2.1 and benchmark MMSE-DFT versus the SNR, under different antenna setups of $N_t \times N_r = 16 \times 16$ and 32×16 with (a) weak channel correlation of $\epsilon_t = 0.3$, and (b) strong channel correlation of $\epsilon_t = 0.8$

2.6 Conclusion

In this chapter, the training optimizations for hybrid MIMO communications have been investigated. Different from most of the existing works that rely on the existence of some special structures in the channel matrix to simplify channel estimation, we have considered a more general channel matrix without any special structures. Two-channel estimation schemes have been proposed for hybrid MIMO systems. In each scheme, the training sequence and the analog matrices at the receiver and transmitter have been optimized for both sum MSE minimization and MI maximization. In the first scheme, the optimal structures of the analog matrices have been derived rigorously, which can overcome the nonconvex nature of analog matrix optimizations and simplify the joint optimizations of training sequences and analog matrices simultaneously. For the second scheme, the analog matrices have been optimized based on the statistical optimization framework, and the training sequence has been effectively optimized according to different channel statistical information. Simulation results have been used to demonstrate the effectiveness of the proposed designs.

Appendix 1

Consider the following statistical optimization

$$\max\ \mathbb{E}\left\{f\left(\boldsymbol{\Psi}^{\mathrm{H}}\boldsymbol{H}_{\mathrm{W}}^{\mathrm{H}}\boldsymbol{\Sigma}^{\mathrm{H}}\boldsymbol{\Sigma}\boldsymbol{H}_{\mathrm{W}}\boldsymbol{\Psi}\right)\right\}, \tag{2.88}$$

where the elements of $\boldsymbol{H}_{\mathrm{W}}$ are i.i.d. Gaussian distributed random variables. The objective function is expressed by

$$\mathbb{E}\left\{f\left(\boldsymbol{\Psi}^{\mathrm{H}}\boldsymbol{H}_{\mathrm{W}}^{\mathrm{H}}\boldsymbol{\Sigma}^{\mathrm{H}}\boldsymbol{\Sigma}\boldsymbol{H}_{\mathrm{W}}\boldsymbol{\Psi}\right)\right\} = \int f\left(\boldsymbol{\Psi}^{\mathrm{H}}\boldsymbol{H}_{\mathrm{W}}^{\mathrm{H}}\boldsymbol{\Sigma}^{\mathrm{H}}\boldsymbol{\Sigma}\boldsymbol{H}_{\mathrm{W}}\boldsymbol{\Psi}\right)p(\boldsymbol{H}_{\mathrm{W}})\mathrm{d}\boldsymbol{H}_{\mathrm{W}}, \tag{2.89}$$

where $p(\boldsymbol{H}_{\mathrm{W}})$ is the probability density function of $\boldsymbol{H}_{\mathrm{W}}$, and $f(\cdot)$ is a unitary invariant function, i.e.,

$$f(\boldsymbol{U}_{\mathrm{L}}\boldsymbol{\Psi}^{\mathrm{H}}\boldsymbol{H}_{\mathrm{W}}^{\mathrm{H}}\boldsymbol{\Sigma}^{\mathrm{H}}\boldsymbol{\Sigma}\boldsymbol{H}_{\mathrm{W}}\boldsymbol{\Psi}\boldsymbol{U}_{\mathrm{L}}^{\mathrm{H}}) = f(\boldsymbol{\Psi}^{\mathrm{H}}\boldsymbol{H}_{\mathrm{W}}^{\mathrm{H}}\boldsymbol{\Sigma}^{\mathrm{H}}\boldsymbol{\Sigma}\boldsymbol{H}_{\mathrm{W}}\boldsymbol{\Psi}), \tag{2.90}$$

for any unitary matrix $\boldsymbol{U}_{\mathrm{L}}$ of appropriate dimension. In addition, $f(\cdot)$ is a matrix-monotone increasing function, i.e., for two positive semidefinite matrices $\boldsymbol{A} \succeq \boldsymbol{B}$, $f(\boldsymbol{A}) \geq f(\boldsymbol{B})$.

Conclution 2.5 *The optimization problem (2.88) in nature aims to maximize* $\boldsymbol{\lambda}(\boldsymbol{\Sigma}^{\mathrm{H}}\boldsymbol{\Sigma})$ *and* $\boldsymbol{\lambda}(\boldsymbol{\Psi}\boldsymbol{\Psi}^{\mathrm{H}})$.

Proof For $\boldsymbol{\Sigma}_1^{\mathrm{H}}\boldsymbol{\Sigma}_1 \succeq \boldsymbol{\Sigma}_2^{\mathrm{H}}\boldsymbol{\Sigma}_2$, we have

$$f(\boldsymbol{\Psi}^{\mathrm{H}}\boldsymbol{H}_{\mathrm{W}}^{\mathrm{H}}\boldsymbol{\Sigma}_1^{\mathrm{H}}\boldsymbol{\Sigma}_1\boldsymbol{H}_{\mathrm{W}}\boldsymbol{\Psi})p(\boldsymbol{H}_{\mathrm{W}})\mathrm{d}\boldsymbol{H}_{\mathrm{W}} \geq f(\boldsymbol{\Psi}^{\mathrm{H}}\boldsymbol{H}_{\mathrm{W}}^{\mathrm{H}}\boldsymbol{\Sigma}_2^{\mathrm{H}}\boldsymbol{\Sigma}_2\boldsymbol{H}_{\mathrm{W}}\boldsymbol{\Psi})p(\boldsymbol{H}_{\mathrm{W}})\mathrm{d}\boldsymbol{H}_{\mathrm{W}}, \tag{2.91}$$

based on which the following inequality holds

$$\mathbb{E}\{f(\boldsymbol{\Psi}^{\mathrm{H}}\boldsymbol{H}_{\mathrm{W}}^{\mathrm{H}}\boldsymbol{\Sigma}_1^{\mathrm{H}}\boldsymbol{\Sigma}_1\boldsymbol{H}_{\mathrm{W}}\boldsymbol{\Psi})\} \geq \mathbb{E}\{f(\boldsymbol{\Psi}^{\mathrm{H}}\boldsymbol{H}_{\mathrm{W}}^{\mathrm{H}}\boldsymbol{\Sigma}_1^{\mathrm{H}}\boldsymbol{\Sigma}_1\boldsymbol{H}_{\mathrm{W}}\boldsymbol{\Psi})\}. \tag{2.92}$$

Moreover, for $\boldsymbol{\lambda}(\boldsymbol{\Sigma}_1^{\mathrm{H}}\boldsymbol{\Sigma}_1) \succeq \boldsymbol{\lambda}(\boldsymbol{\Sigma}_2^{\mathrm{H}}\boldsymbol{\Sigma}_2)$, there always exists a unitary matrix $\boldsymbol{U}$ that makes $\boldsymbol{U}\boldsymbol{\Sigma}_1^{\mathrm{H}}\boldsymbol{\Sigma}_1\boldsymbol{U}^{\mathrm{H}} \succeq \boldsymbol{\Sigma}_2^{\mathrm{H}}\boldsymbol{\Sigma}_2$. Therefore,

$$\begin{aligned}
&\int f(\boldsymbol{\Psi}^{\mathrm{H}}\boldsymbol{H}_{\mathrm{W}}^{\mathrm{H}}\boldsymbol{\Sigma}_1^{\mathrm{H}}\boldsymbol{\Sigma}_1\boldsymbol{H}_{\mathrm{W}}\boldsymbol{\Psi})p(\boldsymbol{H}_{\mathrm{W}})\mathrm{d}\boldsymbol{H}_{\mathrm{W}}\\
&=\int f(\boldsymbol{\Psi}^{\mathrm{H}}\boldsymbol{H}_{\mathrm{W}}^{\mathrm{H}}\boldsymbol{U}^{\mathrm{H}}\boldsymbol{U}\boldsymbol{\Sigma}_1^{\mathrm{H}}\boldsymbol{\Sigma}_1\boldsymbol{U}^{\mathrm{H}}\boldsymbol{U}\boldsymbol{H}_{\mathrm{W}}\boldsymbol{\Psi})p(\boldsymbol{H}_{\mathrm{W}})\mathrm{d}\boldsymbol{H}_{\mathrm{W}}\\
&=\int f(\boldsymbol{\Psi}^{\mathrm{H}}\boldsymbol{H}_{\mathrm{W}}^{\mathrm{H}}\boldsymbol{U}^{\mathrm{H}}\boldsymbol{U}\boldsymbol{\Sigma}_1^{\mathrm{H}}\boldsymbol{\Sigma}_1\boldsymbol{U}^{\mathrm{H}}\boldsymbol{U}\boldsymbol{H}_{\mathrm{W}}\boldsymbol{\Psi})p(\boldsymbol{U}\boldsymbol{H}_{\mathrm{W}})\mathrm{d}\boldsymbol{U}\boldsymbol{H}_{\mathrm{W}}\\
&=\int f(\boldsymbol{\Psi}^{\mathrm{H}}\boldsymbol{H}_{\mathrm{W}}^{\mathrm{H}}\boldsymbol{U}\boldsymbol{\Sigma}_1^{\mathrm{H}}\boldsymbol{\Sigma}_1\boldsymbol{U}^{\mathrm{H}}\boldsymbol{H}_{\mathrm{W}}\boldsymbol{\Psi})p(\boldsymbol{H}_{\mathrm{W}})\mathrm{d}\boldsymbol{H}_{\mathrm{W}}\\
&\geq\int f(\boldsymbol{\Psi}^{\mathrm{H}}\boldsymbol{H}_{\mathrm{W}}^{\mathrm{H}}\boldsymbol{\Sigma}_2^{\mathrm{H}}\boldsymbol{\Sigma}_2\boldsymbol{H}_{\mathrm{W}}\boldsymbol{\Psi})p(\boldsymbol{H}_{\mathrm{W}})\mathrm{d}\boldsymbol{H}_{\mathrm{W}},
\end{aligned} \tag{2.93}$$

where the second and the third equalities are due to the facts that the elements of $\boldsymbol{H}_{\mathrm{W}}$ are i.i.d. Gaussian distributed and $\boldsymbol{H}_{\mathrm{W}}$ and $\boldsymbol{U}\boldsymbol{H}_{\mathrm{W}}$ have the same distribution. The final inequality is derived based on (2.91).

Note that $\boldsymbol{\Psi}^{\mathrm{H}}\boldsymbol{H}_{\mathrm{W}}^{\mathrm{H}}\boldsymbol{\Sigma}^{\mathrm{H}}\boldsymbol{\Sigma}\boldsymbol{H}_{\mathrm{W}}\boldsymbol{\Psi}$ and $\boldsymbol{\Sigma}\boldsymbol{H}_{\mathrm{W}}\boldsymbol{\Psi}\boldsymbol{\Psi}^{\mathrm{H}}\boldsymbol{H}_{\mathrm{W}}^{\mathrm{H}}\boldsymbol{\Sigma}^{\mathrm{H}}$ have the same nonzero eigenvalues. As a result, the following two optimization problems are equivalent

$$\max \boldsymbol{\lambda}(\boldsymbol{\Psi}^{\mathrm{H}}\boldsymbol{H}_{\mathrm{W}}^{\mathrm{H}}\boldsymbol{\Sigma}^{\mathrm{H}}\boldsymbol{\Sigma}\boldsymbol{H}_{\mathrm{W}}\boldsymbol{\Psi}) \Leftrightarrow \max \boldsymbol{\lambda}(\boldsymbol{\Sigma}\boldsymbol{H}_{\mathrm{W}}\boldsymbol{\Psi}\boldsymbol{\Psi}^{\mathrm{H}}\boldsymbol{H}_{\mathrm{W}}^{\mathrm{H}}\boldsymbol{\Sigma}^{\mathrm{H}}). \tag{2.94}$$

Based on the unitary invariant property in (2.90) and (2.93), it can be proved that when $\boldsymbol{\lambda}(\boldsymbol{\Psi}_1\boldsymbol{\Psi}_1^{\mathrm{H}}) \succeq \boldsymbol{\lambda}(\boldsymbol{\Psi}_2\boldsymbol{\Psi}_2^{\mathrm{H}})$, the following inequality holds

$$\int f(\boldsymbol{\Psi}_1^{\mathrm{H}}\boldsymbol{H}_{\mathrm{W}}^{\mathrm{H}}\boldsymbol{\Sigma}_1^{\mathrm{H}}\boldsymbol{\Sigma}_1\boldsymbol{H}_{\mathrm{W}}\boldsymbol{\Psi}_1)p(\boldsymbol{H}_{\mathrm{W}})\mathrm{d}\boldsymbol{H}_{\mathrm{W}} \geq \int f(\boldsymbol{\Psi}_2^{\mathrm{H}}\boldsymbol{H}_{\mathrm{W}}^{\mathrm{H}}\boldsymbol{\Sigma}_1^{\mathrm{H}}\boldsymbol{\Sigma}_1\boldsymbol{H}_{\mathrm{W}}\boldsymbol{\Psi}_2)p(\boldsymbol{H}_{\mathrm{W}})\mathrm{d}\boldsymbol{H}_{\mathrm{W}}. \tag{2.95}$$

This completes the proof.

References

1. Zhang P, Chen S, Hanzo L (2015) Two-tier channel estimation aided near-capcity MIMO transceivers relaying on norm-based joint transmit and receive antenna selection. IEEE Trans Wirel Commun 14(1):122–137
2. Wang Z, Liu W, Qian C et al (2017) Two-dimensional precoding scheme for 3-D massive MIMO. IEEE Trans Veh Technol 66(6):5488–5493
3. Zhang J, Chen S, Maunder RG et al (2018) Adaptive coding and modulation for large-scale antenna array-based aeronautical communications in the presence of co-channel interference. IEEE Trans Wirel Commun 17(2):1343–1357
4. Liu W, Wang Z, Cao J et al (2018) Partially-activated conjugate beamforming for LoS massive MIMO communications. IEEE Access 6:56504–56513
5. Pastore A, Joham M, Fonollosa JR (2016) A framework for joint design of pilot sequence and linear precoder. IEEE Trans Inf Theory 62(9):5059–5079
6. Hassibi B, Hochwald BM (2003) How much training is needed in multiple-antenna wireless links? IEEE Trans Inf Theory 49(4):951–963
7. Soysal A, Ulukus S (2010) Joint channel estimation and resource allocation for MIMO systems-Part I: single-user analysis. IEEE Trans Wirel Commun 9(2):624–631
8. Soysal A, Ulukus S (2010) Joint channel estimation and resource allocation for MIMO systems-part II: multi-user and numerical analysis. IEEE Trans Wirel Commun 9(2):632–640
9. Coldrey M, Bohlin P (2007) Training-based MIMO systems-Part I: performance comparison. IEEE Trans Signal Process 55(11):5464–5476
10. Coldrey M, Bohlin P (2008) Training-based MIMO systems: part II-Improvements using detected symbol information. IEEE Trans Signal Process 56(1):296–303
11. Gao Z, Dai L, Wang Z et al (2015) Spatially common sparsity based adaptive channel estimation and feedback for FDD massive MIMO. IEEE Trans Signal Process 63(23):6169–6183
12. Wong TF, Park B (2004) Training sequence optimization in MIMO systems with colored interference. IEEE Trans Commun 52(11):1939–1947
13. Liu Y, Wong TF, Hager WW (2007) Training signal design for estimation of correlated MIMO channels with colored interference. IEEE Trans Signal Process 55(4):1486–1497
14. Biguesh M, Gershman AB (2006) Training-based MIMO channel estimation: a study of estimator tradeoffs and optimal training signals. IEEE Trans Signal Process 54(3):884–893
15. Gao F, Cui T, Nallanathan A (2008) Optimal training design for channel estimation in decode-and-forward relay networks with individual and total power constraints. IEEE Trans Signal Process 56(12):5937–5949
16. Guo X, Chen S, Zhang J et al (2016) Optimal pilot design for pilot contamination elimination/reduction in large-scale multiple-antenna aided OFDM systems. IEEE Trans Wirel Commun 15(11):7229–7243
17. Xie H, Gao F, Jin S et al (2018) Channel estimation for TDD/FDD massive MIMO systems with channel covariance computing. IEEE Trans Wirel Commun 17(6):4206–4218
18. Cheng X, Sun J, Li S (2017) Channel estimation for FDD multi-User massive MIMO: a variational Bayesian inference-based approach. IEEE Trans Wirel Commun 16(11):7590–7602
19. Wu C-Y, Huang W-J, Chung W-H (2017) Low-complexity semiblind channel estimation in massive MU-MIMO systems. IEEE Trans Wirel Commun 16(9):6279–6290
20. Xiong X, Wang X, Gao X et al (2017) Beam-domain channel estimation for FDD massive MIMO systems with optimal thresholds. IEEE Trans Wirel Commun 16(7):4669–4682
21. Fang J, Li X, Li H et al (2017) Low-rank covariance-assisted downlink training and channel estimation for FDD massive MIMO systems. IEEE Trans Wirel Commun 16(3):1935–1947
22. Fan D, Gao F, Wang G et al (2017) Angle domain signal processing-aided channel estimation for indoor 60-GHz TDD/FDD massive MIMO systems. IEEE J Sel Areas Commun 35(9):1948–1961
23. Zhao J, Gao F, Jia W et al (2017) Angle domain hybrid precoding and channel tracking for millimeter wave massive MIMO systems. IEEE Trans Wirel Commun 16(10):6868–6880

24. Xu W, Xiang W, Jia Y et al (2017) Downlink performance of massive-MIMO systems using EVD-based channel estimation. IEEE Trans Veh Technol 66(4):3045–3058
25. Kong D, Qu D, Luo K et al (2016) Channel estimation under staggered frame structure for massive MIMO system. IEEE Trans Wirel Commun 15(2):1469–1479
26. Xing C, Zhao X, Xu W et al (2019) A framework on hybrid MIMO transceiver design based on matrix-monotonic optimization. IEEE Trans Signal Process 67(13):3531–3546
27. Pan L, Liang L, Xu W et al (2018) Framework of channel estimation for hybrid analog-and-digital processing enabled massive MIMO communications. IEEE Trans Commun 66(9):3902–3915
28. Ni W, Dong X (2016) Hybrid block diagonalization for massive multiuser MIMO systems. IEEE Trans Commun 64(1):201–211
29. Ni W, Dong X, Lu W-S (2017) Near-optimal hybrid processing for massive MIMO systems via matrix decomposition. IEEE Trans Signal Process 65(15):3922–3933
30. Xing C, Ma S, Zhou Y (2015) Matrix-monotonic optimization for MIMO systems. IEEE Trans Signal Process 63(2):334–348
31. Katselis D, Kofidis E, Theodoridis S (2008) On training optimization for estimation of correlated MIMO channels in the presence of multiuser interference. IEEE Trans Signal Process 56(10):4892–4904
32. Kay S (1993) Fundamental of statistical signal processing: estimation theory. Prentice-Hall
33. Marshall AW, Olkin I (1979) Inequalities: theory of majorization and its applications. Academic Press
34. Boyd S, Vandenberghe L (2004) Convex optimization. Cambridge University Press
35. Jin J, Xiao C, Chen W et al (2019) Channel-statistics-based hybrid precoding for millimeter-wave MIMO systems with dynamic subarrays. IEEE Trans Commun 67(6):3991–4003
36. Liu A, Lau VKN (2016) Impact of CSI knowledge on the codebook-based hybrid beamforming in massive MIMO. IEEE Trans Signal Process 64(24):6545–6556
37. Zhu D, Li B, Liang P (2017) A novel hybrid beamforming algorithm with unified analog beamforming by subspace construction based on partial CSI for massive MIMO-OFDM systems. IEEE Trans Commun 65(2):594–607

Chapter 3
Accurate Delay Estimation in MIMO Communication Systems

Deguang Zhao, Chen Zheng, and Xuhui Ding

This chapter gives a time delay estimation method based on the spectrum structure of constant envelope signal and proves that the estimation method can gradually reach the CRLB. Section 3.1 introduces the research status of accurate estimation of time delay in MIMO communication systems. Section 3.2 describes the system structure model of the terahertz (THz) large-scale antenna array and gives the received signal model. Section 3.3 derives the CRLB based on the time delay estimation of the system model. An estimation method for progressively reaching the CRLB is given in this section. In Sect. 3.4, the optimal training sequence design and optimal TDL matrix design methods are given with the goal of minimizing the CRLB. Simulation results are given in Sect. 3.5 to evaluate the performance of proposed the algorithm. Conclusions can be found in Sect. 3.6.

3.1 Introduction

Multi-antenna array is a key technology that is capable of both making full use of spatial multiplexing and obtaining diversity gain in wireless communication systems [1–3]. It has been shown that future high-speed data networks need to rely on massive MIMO technology. Furthermore, for some special scenarios, the physical size constraints of the communication devices are no longer as strict as those in

D. Zhao · C. Zheng · X. Ding (✉)
Beijing Institute of Technology, Beijing, China
e-mail: dingxuhui@bit.edu.cn

D. Zhao
e-mail: zhaodeguang@bit.edu.cn

C. Zheng
e-mail: mz3721@126.com

X. Ding et al. (eds.), *Performance Analysis and Improvement in MIMO Communication Systems*, https://doi.org/10.1007/978-981-96-5951-7_3

the cellular networks, allowing the size of the antenna arrays to become relatively larger. With the increasing requirements of channel capacity, the bandwidth of the traditional communication system can no longer meet the corresponding requirements [4, 5]. Fortunately, the THz technology enables us to utilize a larger bandwidth, a shorter wavelength, and a smaller antenna size, thus allowing us to carry more antennas in space-limited scenarios. Meanwhile, spatially dispersed users can be supported simultaneously due to the high-frequency band, weak signal scatters, and the channel's sparsity characters of the THz system.

Although THz technology has many advantages, some technical challenges still remain. Among them, synchronization is the core functionality of communication systems, especially for the THz high-speed communication system [6]. The large bandwidth characteristic of the signal makes it difficult for the analog-to-digital converter (ADC) as well as digital-to-analog converter (DAC) to reach the Nyquist sampling rate while maintaining a high quantization bit number [7, 8]. Furthermore, the accuracy of both ADC and DAC deteriorates drastically at very high sampling rates, making them unsuitable for high-order modulation. Under this condition, only simple modulation methods, such as binary phase shift keying (BPSK) and on-off keying (OOK) modulation types can be realized [9]. Meanwhile, extremely high accuracy in delay estimation is demanded due to ultrashort symbol duration. Compared with the orthogonal frequency division multiplexing (OFDM) method, the accuracy of delay estimation in single-carrier modulation must be well within the interval of one sampling clock [10–12]. Basically, even tiny errors can significantly degrade the performance of the THz systems [13, 14].

Generally, the output power of the THz amplifier is limited, and the THz signal attenuates more rapidly with distance. Therefore, to guarantee sufficient link margin, the corresponding antenna beam width should be narrow, which requires an extremely high AoA estimation accuracy [15–17]. Under this condition, even a small azimuth angle error will cause a great loss of SNR and then destroy the quality of communication. Meanwhile, due to the real-time requirements of application scenarios, angle estimation needs to be completed in a very short time such as the unmanned aerial vehicle (UAV) that has a line-of-sight (LoS) channel with users [18–20]. This undoubtedly requires a high-precision, high-speed estimation method [17, 21–24].

For broadband array models, traditional research has focused on wideband direction of arrival (DoA) estimation and wideband beamforming to optimize the channel capacity. The corresponding method is to maximize the received SNR or minimize the signal estimation variance [21, 22]. However, in the broadband array, maximizing the SNR might not guarantee the optimal delay estimation accuracy. Furthermore, the huge computation complexity caused by the large array and wideband signal is another difficulty. To reduce the computational complexity, one solution is to simplify the complexity by using sparsity of angle, such as sparse array, analog matrix [25–30], and another method is to simplify the delay estimator by using sparsity of time delay, such as wideband modulation converter (WMC) structure [31–34]. A novel method named generalized approximated message passing (GAMP) based CS recovery technology is an attracting interest [35–40]. However, the method based on

CS inevitably has the problem of base mismatch, thereby a high-precision estimation method is necessary.

3.2 System Structure and Signal Model

Due to the immature process of THz components, the output power is relatively small. In order to utilize the efficiency of the power amplifier at the transmitting end as much as possible, the constant envelope training pilot is used in this chapter. The constant modulus phase-coded training pilot containing I symbols is given by

$$x(t) = \sum_{i=0}^{I-1} x_i \cdot p(t - iT_f), \tag{3.1}$$

where $x_i = \rho e^{(j\theta_i)}$ represents the i-th training symbol. To guarantee constant modulus, $\rho = 1$. $p(t)$ representing a shaped pulse for one symbol, could be chosen as a rectangular pulse, a root-raised-cosine pulse, or a Gaussian pulse. The duration of one symbol is T_f. The frequency-domain response of this pilot is given by

$$X(f) = FT\{x(t)\} = \sum_{i=0}^{I-1} x_i \cdot P(f) \cdot e^{j2\pi f(i \cdot T_f)}. \tag{3.2}$$

The THz channel response is $h(t - \tau)$, and the frequency response is $H(f)e^{j2\pi f\tau}$, correspondingly. Specifically,

$$h(t) = F^{-1}\{H(f)\}, \tag{3.3}$$

$$H(f) = \frac{c}{4\pi f d_T} \cdot \exp\left(-1/2\, K(f) d_T\right), \tag{3.4}$$

where $K(f)$ represents the attenuation coefficient in THz frequency f. d_T represents the distance between the base station (BS) and the user. And τ represents the time delay from BS to the user, which we need to estimate. The signal received through the THz channel and its spectrum can be deduced as follows

$$y(t) = \sum_{i=0}^{I-1} x_i \cdot g(t - iT_f - \tau), \tag{3.5}$$

$$Y(f) = \sum_{i=0}^{I-1} x_i \cdot G(f) \cdot e^{j2\pi f(iT_f)} \cdot e^{j2\pi f\tau}. \tag{3.6}$$

Here, $g(t) = p(t) * h(t)$ represents the pulse after the shaped pulse passes through the THz channel. The Fourier transform is $G(f)$. Considering that the reflection of the THz signal is very weak in actual conditions, the power of such a multipath signal is much smaller than that of LoS. Therefore, only the signal of the direct path is considered in this chapter. At the BS side, consider a line array receiver with M antennas. The interval between two adjacent antennas is d. Thereby the dimension of this antenna array is $(M-1)d$. For example, a THz communication system occupies center frequency $f_c = 150$ gigahertz (GHz) which wavelength $\lambda = c/f = 2$ mm. The interval between two antenna is $d = \lambda/2 = 1$ mm. So, in a 128 antennas array, the size is $D = (M-1)d = 127$ mm. Meanwhile, a signal with 10 GHz symbol rate is transmitted, which duration $T_f = 100$ ps. Then,

$$T_f \times c = 100\,\text{ps} \times 3 \times 10^8\,\text{m/s} = 30\,\text{mm} \leq 127\,\text{mm}. \tag{3.7}$$

On such occasions, when the first symbol has just arrived at the No. 0 antenna, the fifth symbol is already on the No. 127 antenna, where the existing narrowband assumption is no longer satisfied. At this time, each antenna is no longer a different phase-shifted copy of the same symbol, but a completely different symbol. In this case, we need to reintroduce the model of the received signal from the wideband array. Consider the signal arrived at m-th antenna is $y_m(t)$ with AoA θ. $y_m(t)$ can be derived as follows

$$y_m(t) = y(t - m\frac{d\sin\theta}{c}) \times e^{j2\pi f_c m \frac{d\sin\theta}{c}}. \tag{3.8}$$

Denote $\Delta\tau = \frac{d\sin\theta}{c}$ as the time delay between signal reached two adjacent antennas. We can get

$$y_m(t) = y(t - m\Delta\tau) = \sum_{i=0}^{I-1} x_i \cdot g(t - iT_f - \tau - m\Delta\tau)e^{j2\pi f_c m\Delta\tau}. \tag{3.9}$$

Correspondingly, we have

$$Y_m(f) = Y(f)e^{j2\pi f m\Delta\tau}e^{j2\pi f_c m\Delta\tau}. \tag{3.10}$$

For the size $(M-1)d$ of the antenna array, it is much larger than the length $(T_f \times C)$ of a symbol propagating in space, so we employed an array receiver structure called a TDL. Each antenna is followed by a set of time-delay lines (increased literature), each of which delays the signal by ΔT, and a series of delayed signals are summed by the digital phase shifter. Traditional TDL arrays require both digital phase shifters and digitally controlled attenuators to control the amplitude and phase of the signal. Intuitively, the time delay structure can align the signals on different antennas in time and then combine them, but this requires the coefficients of the TDL structure to have a two-dimensional adjustment of phase and amplitude, which is not suitable for large-scale antenna arrays. Therefore, in order to simplify the design, this chapter uses a phase-shifting structure array to only phase shift the signal delay. Due to the constant mode structure, it is not suitable for analysis in the time domain and needs to be considered from the frequency point of view.

Thus, the signal $r(t)$ passing through the TDL and digital phase shifters can be written as the summation of the M antenna branch signals after passing through the TDL structure.

$$r(t) = \sum_{m=0}^{M-1} r_m(t). \tag{3.11}$$

In which, $r_m(t)$ represents the summation of the signal received in the m-th antenna passing through TDL structure.

$$r_m(t) = \sum_{n=0}^{N-1} [y_m(t - n\Delta T) + n_m(t - n\Delta T)]\, w_{m,n}. \tag{3.12}$$

Substitute $r_m(t)$ into the summation, we have

$$\begin{aligned} r(t) =& \sum_{m=0}^{M-1}\sum_{n=0}^{N-1} y(t - m\Delta\tau - n\Delta T)e^{j2\pi f_c m\Delta\tau} w_{m,n} \\ &+ \sum_{m=0}^{M-1}\sum_{n=0}^{N-1} n_m(t - n\Delta T) w_{m,n}. \end{aligned} \tag{3.13}$$

Here, $R(f)$ is given by

$$\begin{aligned} R(f) =& Y(f) \sum_{m=0}^{M-1} \left[e^{j2\pi(fm\Delta\tau + fn\Delta T + f_c m\Delta\tau)}\right] w_{m,n} \\ &+ \sum_{m=0}^{M-1} N_m(f) \left[e^{j2\pi f n\Delta T} w_{m,n}\right]. \end{aligned} \tag{3.14}$$

The $r(t)$ is sampled in the time domain, with the sampling interval T_s, and the sampled signal $z(n)$. Its Fourier transform $Z(k)$ can be written in the following form.

$$\begin{aligned} z(n) =& r(nT_s) \\ =& \sum_{m=0}^{M-1}\sum_{n=0}^{N-1} y(nT_s - m\Delta\tau - n\Delta T)e^{j2\pi f_c m\Delta\tau} w_{m,n} \\ &+ \sum_{m=0}^{M-1}\sum_{n=0}^{N-1} n_m(nT_s - n\Delta T) w_{m,n}. \end{aligned} \tag{3.15}$$

$$\begin{aligned} Z(k) =& Y(f) \sum_{m=0}^{M-1} \sum_{n=0}^{N-1} \left[e^{j2\pi(fm\Delta\tau + fn\Delta T + f_c m\Delta\tau)} \right] w_{m,n} \\ &+ \sum_{m=0}^{M-1} N_m(f) \left[e^{j2\pi fn\Delta T} w_{m,n} \right] \\ =& Y(f_k) Tr(\boldsymbol{E}(f_k)\boldsymbol{W}^T) + \boldsymbol{n}(f_k)^T \boldsymbol{W} \boldsymbol{e}_n(f_k). \end{aligned} \tag{3.16}$$

In which,

$$\boldsymbol{E}(f_k)_{m,n} = \left[e^{j2\pi(fm\Delta\tau + fn\Delta T + f_c m\Delta\tau)} \right], \tag{3.17}$$

$$\boldsymbol{e}_n(f_k) = \left[e^{j2\pi f(0)\Delta T}, \ldots, e^{j2\pi f(N-1)\Delta T} \right]^T, \tag{3.18}$$

$$\boldsymbol{n}(f_k) = \left[N_0(f_k), \ldots, N_{M-1}(f_k) \right]^T. \tag{3.19}$$

The essence of TDL is the weighted summation of the $M \times N$ different delayed signals $y(t - m\Delta\tau - n\Delta T)$ of the received signal $y(t - \tau)$. In this chapter, to reduce the system cost, the entire TDL array finally synthesizes a signal $r(t)$ corresponding to one RF link. For the sake of generality, it is considered to increase the number of RF links, that is, both the digital-analog mixing and the TDL structure.

3.3 Near CRLB Estimation Method of Time Delay

According to the expression form of $Z(k)$, it can be seen that the spectrum of the received signal contains the following components:

$$\begin{aligned} Z(k) =& X(f_k) H(f_k) e^{j2\pi f_k \tau} \times Tr(\boldsymbol{E}(f_k)\boldsymbol{W}^T) + \boldsymbol{n}(f_k)^T \boldsymbol{W} \boldsymbol{e}_n(f_k) \\ =& \left[\sum_{i=0}^{I-1} x_i G(f_k) e^{j2\pi f(iT_f)} \right] e^{j2\pi f_k \tau} \\ &\times \sum_{m=0}^{M-1} \sum_{n=0}^{N-1} \left[e^{j2\pi(fm\Delta\tau + fn\Delta T + f_c m\Delta\tau)} \right] \times w_{m,n} \\ &+ \sum_{m=0}^{M-1} N_m(f) \left[e^{j2\pi fn\Delta T} \times w_{m,n} \right]. \end{aligned} \tag{3.20}$$

- Training pilot: $X(f_k) = \sum_{i=0}^{I-1} x_i G(f_k) e^{j2\pi f(iT_f)}$.
- Time delay: $e^{j2\pi f_k \tau}$.
- Angle of arrival: $e^{j2\pi fm\Delta\tau} e^{j2\pi f_c m\Delta\tau}$. $e^{j2\pi fm\Delta\tau}$ is caused by the too-short duration of a pulse, which is the biggest difference between traditional narrowband arrays. $e^{j2\pi f_c m\Delta\tau}$ is caused by the center frequency, which is the same as the traditional narrowband array.

- Time delay line: $e^{j2\pi f n\Delta T}$. This is unique in TDL. Different from the existing analog-digital hybrid arrays, thanks to this item, signals from different antennas can be aligned in the time domain.
- Phase shifter matrix: $w_{m,n}$. By properly designing this matrix, signals can be summed in phase, which could increase the receiving signal-noise ratio.
- Frequency domain noise: $n(f_k)$. because of the additive white Gaussian noise (AWGN) assumption, noises in M antenna with different frequencies are independent and identically distributed.

Based on previous analysis, in view of the broadband of signal and the huge dimension of arrays, for a specific $\Delta\tau$, $f_k m\Delta\tau$ could not be negligible, and $e^{j2\pi f_k m\Delta\tau}$ can't be approximated by 1. Thus, we need to rewrite $Z(k)$ as follows

$$Z(k) = X(f_k) \cdot A(f_k, \Delta\tau) \cdot e^{j2\pi f_k \tau} + W(f_k). \tag{3.21}$$

In which, $A(f_k, \Delta\tau)$ represents the inference arising from space angle. $e^{j2\pi f_k \tau}$ contains the time delay which we need to estimate. $A(f_k, \Delta\tau)$ is given by

$$A(f_k, \Delta\tau) = \sum_{m=0}^{M-1}\sum_{n=0}^{N-1}\left[e^{j2\pi(fm\Delta\tau + fn\Delta T + f_c m\Delta\tau)}\right] w_{m,n}. \tag{3.22}$$

There has been a lot of literature on broadband DoA estimation. In this case, we assume $\Delta\tau$ is known. Thus $A(f_k, \Delta\tau)$ is known at the same time, and can be abbreviated to $A(f_k, \Delta\tau) = A(f_k)$. Now, we need to estimate τ.

$$Z(k) = X(f_k)A(f_k)e^{j2\pi f_k \tau} + W(f_k). \tag{3.23}$$

From the expression of $Z(k)$, it can be seen clearly that under zero noise, $Z(k) = X(f_k)A(f_k)e^{j2\pi f_k \tau}$. Estimation of time delay can be replaced by estimating the phase of $Z(k)$ in different frequencies. For a specific frequency f_k, we can find

$$\hat{\tau} = \frac{1}{2\pi f_k}\left(\angle Z(k) - \angle X(f_k) - \angle A(f_k)\right). \tag{3.24}$$

In which, $X(f_k)$ and $A(f_k)$ can be calculated in advance. Accordingly, the estimation of τ follows direct spectrum estimation of $Z(k)$. $Z(k)$ can be simply estimated by fast Fourier transform (FFT) or Wiener-Hopf method.

3.3.1 Deduction of CRLB

According to the expression of signal, we could rewrite it in the following form

$$Z(k) = X(f_k)A(f_k)e^{j2\pi f_k \tau} + W(f_k) = s(k;\tau) + n(k). \tag{3.25}$$

In which, $s(k;\tau) = X(f_k)A(f_k)e^{j2\pi f_k \tau}$. Then we have the likelihood function of the received signal

$$p(Z;\tau) = \prod_{k=0}^{K-1}\left[\frac{1}{\sqrt{2\pi\sigma_k^2}}exp\{-\frac{1}{2\sigma_k^2}(Z(k) - s(k;\tau))^2\}\right]. \tag{3.26}$$

By derivation and performing expectation, we can get

$$E\left(\frac{\partial ln(p(Z;\tau))}{\partial\tau}\right) = -\sum_{k=0}^{K-1} -\frac{1}{\sigma_k^2}\left(\frac{\partial s(k;\tau)}{\partial\tau}\right)^2. \tag{3.27}$$

Thus,

$$var(\hat{\tau}) \geq \left[\sum_{k=0}^{K-1} -\frac{1}{\sigma_k^2}\left(\frac{\partial s(k;\tau)}{\partial\tau}\right)^2\right]^{-1}. \tag{3.28}$$

Further, we perform derivation to $s(k;\tau)$

$$\begin{aligned}\frac{\partial s(k;\tau)}{\partial\tau} &= \frac{\partial\left[X(f_k)A(f_k)e^{j2\pi f_k\tau}\right]}{\partial\tau} \\ &= (j2\pi f_k)X(f_k)A(f_k)e^{j2\pi f_k\tau}.\end{aligned} \tag{3.29}$$

Substituting this result into (3.28), we could have CRLB as follows

$$\begin{aligned}var(\hat{\tau}) &\geq \left[\sum_{k=0}^{K-1} -\frac{1}{\sigma_k^2}\left(\frac{\partial s(k;\tau)}{\partial\tau}\right)^2\right]^{-1} \\ &= \left[\sum_{k=0}^{K-1} -\frac{1}{\sigma_k^2}\|(j2\pi f_k)X(f_k)A(f_k)e^{j2\pi f_k\tau}\|^2\right]^{-1} \\ &= \left(\sum_{k=0}^{K-1}(2\pi f_k)^2\frac{\|X(f_k)A(f_k)\|^2}{\sigma_k^2}\right)^{-1}.\end{aligned} \tag{3.30}$$

3.3.2 Near CRLB Estimation Method

Consider a specific frequency f_k

$$Z(k) = X(f_k)A(f_k)e^{j2\pi f_k\tau} + W(f_k). \tag{3.31}$$

Denote $\hat{\phi}_k = \angle Z(k)$ as the phase of signal in this frequency, denote $\phi = \angle\left[X(f_k)A(f_k)e^{j2\pi f_k\tau}\right]$ as the phase of noiseless signal in this frequency. Thus we can have an equation

$$\hat{\phi}_k = \angle Z(k) = \arctan\left(\frac{|X(f_k)A(f_k)|\sin\phi + \epsilon_s}{|X(f_k)A(f_k)|\cos\phi + \epsilon_s}\right). \tag{3.32}$$

In which, ϵ_s and ϵ_c represent Im$[W(f_k)]$ and Re$[W(f_k)]$ correspondingly. Therefore, we perform the first-order Taylor expansion on the above estimator, without loss of generality, the following results

$$\begin{aligned}\hat{\phi} &= \arctan\frac{\rho\sin\phi + \epsilon_s}{\rho\cos\phi + \epsilon_c} = g(\epsilon_s, \epsilon_c)\\ &\approx g(0,0) + \frac{\partial g(\epsilon_s, \epsilon_c)}{\epsilon_s} + \frac{\partial g(\epsilon_s, \epsilon_c)}{\epsilon_c}\\ &= \phi + \frac{\partial g(\epsilon_s, \epsilon_c)}{\epsilon_s} + \frac{\partial g(\epsilon_s, \epsilon_c)}{\epsilon_c}.\end{aligned} \tag{3.33}$$

Known that the derivation of $\arctan(x)$ is $1/(1+x^2)$, we can have

$$\hat{\phi} \approx \phi + \frac{1}{\rho}(\cos\phi)\,\epsilon_s + \frac{1}{\rho}(\sin\phi)\,\epsilon_c. \tag{3.34}$$

ϵ_s and ϵ_c are i.i.d., sharing same gaussian distribute $N(0,\sigma^2)$. Thus $(\cos\phi)\,\epsilon_s + (\sin\phi)\,\epsilon_c \sim N(0,\sigma^2)$. After all, we can prove that under small noise $\hat{\phi} \approx \phi + \frac{1}{\rho}n$, in which $n \sim N(0,\sigma^2)$. Then for the specific frequency f_k, the estimation of phase can be expressed as follows

$$\hat{\phi}_k \approx 2\pi f_k\tau + \angle X(f_k) + \angle A(f_k) + n(k). \tag{3.35}$$

In which, $n(k) \sim N\left(0, \frac{\sigma_k^2}{|X(f_k)A(f_k)|^2}\right)$. By the estimation of phase in f_k, we can estimate τ_k

$$\hat{\tau}_k = \frac{\hat{\phi}_k}{2\pi f_k} - \frac{\angle X(k)}{2\pi f_k} - \frac{\angle A(k)}{2\pi f_k}, \tag{3.36}$$

$$\hat{\tau}_k \sim N\left(\tau, \frac{1}{(2\pi f_k)^2}\frac{\sigma_k^2}{|X(f_k)A(f_k)|^2}\right). \tag{3.37}$$

The estimation of the time delay can be achieved using any one of the frequency components f_k. But by observation of $\hat{\tau}_k$ probability distribution, we find that for a higher frequency f_k, $\hat{\tau}_k$ has a bigger denominator $(2\pi f_k)^2$, which means a smaller variance of $\hat{\tau}_k$. This shows that at the same SNR, higher frequency components can provide a more accurate frequency estimation. It is easy to understand that the additional phase caused by the high-frequency components is more pronounced during the same time delay. However, in practical systems, to reduce the Nyquist sampling rate, the high-frequency components in the signal are usually suppressed, and the signal energy is mainly concentrated on the low-frequency components. Moreover, considering design errors in realistic systems, high-frequency component signals may be distorted due to group delay and dispersion effects. Therefore, the time delay estimation accuracy can be improved by appropriately selecting the frequency component f_k or by appropriately distributing the signal power on different frequency components. By estimating the time delay of the K frequency components separately, K estimates $\hat{\tau}_k$ are obtained. Then linearly combine the K estimates to get the final time delay estimation τ.

$$\hat{\tau} = \sum_{k=0}^{K-1} \alpha_k \hat{\tau}_k. \tag{3.38}$$

To get an unbiased estimation, $\sum_{k=0}^{K-1} \alpha_k = 1$

$$\hat{\tau} \sim N\left(\tau, \sum_{k=0}^{K-1} \alpha_k^2 \frac{1}{(2\pi f_k)^2} \frac{\sigma_k^2}{|X(f_k)A(f_k)|^2}\right). \tag{3.39}$$

By minimizing variance, we have

$$\begin{aligned} \min_{\alpha_k} \ & \sum_{k=0}^{K-1} \alpha_k^2 \frac{1}{(2\pi f_k)^2} \frac{\sigma_k^2}{|X(f_k)A(f_k)|^2} \\ \text{s.t.} \ & \sum_{k=0}^{K-1} \alpha_k = 1. \end{aligned} \tag{3.40}$$

The minimum variance is

$$\left(\sum_{k=0}^{K-1} \frac{(2\pi f_k)^2 |X(f_k)A(f_k)|^2}{\sigma_k^2}\right)^{-1}. \tag{3.41}$$

The minimum variance is consistent with CRLB.

3.4 Optimization of Delay Estimation Under TDL Structure

Pseudo-random signals are often used as training sequences in engineering. For example, a phase-encoded signal sampled in a radar, a binary pseudo-random sequence in the global positioning system, and the like. Since the pseudo-random sequence has good autocorrelation, the receiver can determine the delay estimate based on the cross-correlation of the received signal with the local replica. However, on the one hand, the pseudo-random sequence itself does not guarantee optimal performance, usually a sequence generated by a structured algorithm, which is difficult to search through exhaustively; on the other hand, since the receiver usually uses matched filtering, signal delay The estimation accuracy is limited by the sampling rate of the ADC, and the matched filtering requires a large number of parallel addition and multiplication operations, and the computational overhead is extremely large. In traditional pseudo-random synchronization signals, the signal energy is mainly concentrated near the baseband, such as the common singular spectrum. According to our previous discussion, the baseband low-frequency signal can provide limited delay estimation accuracy. Therefore, we need to redesign the optimal synchronous waveform according to the CRLB form of the signal.

In addition to the pilot $X(f_k)$, the design of the TDL array coefficients will directly affect the structure of $A(f_k)$, which in turn affects the performance of the delay estimation, so we also need to optimize the TDL array coefficients. The TDL structure of the array is fixed, and needs to be optimized, which is the phase weighting coefficient $w_{m,n} = e^{j\theta_{m,n}}$ after each stage delay. Wherein, the time delay structure of the array is fixed and needs to be optimized, that is, the phase weighting coefficient after each stage delay is different for the conventional narrowband array, because the different phase copies of the same data symbol are received on each antenna, therefore, the phase weighting coefficient is designed to adjust the phase of each signal so that the beam direction of the antenna is aligned with the AoA. For wideband arrays, the symbols on the individual antennas are different because of the short duration of the data symbols and the large size of the antenna array. Thus, the basic idea of design $w_{m,n}$ is to align the originally unsynchronized symbols. An intuitive solution is to delay the signals received by the antennas in the time domain. However, this requires that both the amplitude and phase of $w_{m,n}$ can be adjusted. Another indirect solution is to understand the TDL structure behind the antenna as a filter. By adjusting the tap coefficients of the filter, the gain of different frequency points of the signal is changed, thereby improving the overall delay estimation accuracy. Thus, based on the previous analysis, we designed the training sequence X and the TDL matrix W to minimize CRLB.

3.4.1 Training Design

In a communication system, in order to ensure communication quality and maximize the SNR of the received signal, the selection of W is usually fixed. So, we assume that W is known to optimize X.

Recalling the previous derivation of the CRLB

$$CRLB = \left(\sum_{k=0}^{K-1} \frac{(2\pi f_k)^2 |X(f_k)A(f_k)|^2}{\sigma_k^2} \right)^{-1}. \tag{3.42}$$

We rewrite the previous problem as follows

$$\begin{aligned} &\max_x \sum_{k=0}^{K-1} \frac{(2\pi f_k)^2 |X(f_k)A(f_k)|^2}{\sigma_k^2} \\ &\text{s.t. } |x_i| = 1. \end{aligned} \tag{3.43}$$

Then we express it in detail

$$X(f_k)A(f_k) = G(f_k)[\boldsymbol{x}^T \boldsymbol{F}(f_k)]Tr[\boldsymbol{E}(f_k)^T \boldsymbol{W}] \tag{3.44}$$

and

$$\begin{aligned} \sigma_k^2 &= E\left[\left(\boldsymbol{n}(f_k)^T \boldsymbol{W} \boldsymbol{e}_n(f_k)\right)^H \left(\boldsymbol{n}(f_k)^T \boldsymbol{W} \boldsymbol{e}_n(f_k)\right) \right] \\ &= \sigma_n^2 Tr\left[\boldsymbol{W} \boldsymbol{e}_n(f_k) \boldsymbol{e}_n(f_k)^H \boldsymbol{W}^H \right]. \end{aligned} \tag{3.45}$$

Then, we have

$$\sum_{k=0}^{K-1} \frac{(2\pi f_k)^2 |X(f_k)A(f_k)|^2}{\sigma_k^2} = \boldsymbol{x}^H \boldsymbol{\Sigma} \boldsymbol{x}. \tag{3.46}$$

In which,

$$\boldsymbol{\Sigma} = \sum_{k=0}^{K-1} \left[(2\pi f_k)^2 \boldsymbol{F}(f_k) \boldsymbol{F}(f_k^H) \frac{|G(f_k)Tr[\boldsymbol{E}(f_k)^T \boldsymbol{W}]|^2}{\sigma_k^2} \right]. \tag{3.47}$$

Therefore, the initial optimization can be converted into such a formation

$$\begin{aligned} &\max_x \boldsymbol{x}^H \boldsymbol{\Sigma} \boldsymbol{x} \\ &\text{s.t. } |x_i| = 1. \end{aligned} \tag{3.48}$$

According to the form of the problem, the feature decomposition can be directly performed on $\boldsymbol{\Sigma}$.

$$\boldsymbol{\Sigma} = \boldsymbol{U} \boldsymbol{\Lambda}_X \boldsymbol{U}^H. \tag{3.49}$$

Denote u as the primitive eigenvector of $\boldsymbol{\Sigma}$. To satisfy the constant modulus constraint, we choose the phase of u as optimal x. Thus, the optimal pilot is given by

$$\boldsymbol{x}^{opt} = exp(j \cdot \angle \boldsymbol{u}). \tag{3.50}$$

3.4.2 TDL Matrix Optimization

For radar and other systems, the radar waveform is usually given. Here, the delay estimation is the most important part. Therefore, the W matrix of the TDL structure needs to be designed. We assume that the training sequence is fixed and only optimized for W. By observing the structure of the system, we can find that $e^{j2\pi fn\Delta T}$ and $w_{m,n}$ together consist of a frequency selection structure, which changes the power corresponding to a different frequency, and thus changes the accuracy of time delay estimation using the line. Under the premise that the incident angle is known, by changing W, the gain of the array for different frequencies can be changed. Intuitively, for frequency $f_k = f_0$ the optimal choice of the coefficient is $w_{m,n}^{opt} = (e^{-j2\pi fm\Delta\tau}e^{-j2\pi fn\Delta T}e^{-j2\pi f_c m\Delta\tau})$. While, for another frequency $f_k = f_1$, this optimal solution may not be applicable at all. So it is necessary to optimize the design. Similar to the design of the training sequence, we first rewrite the original problem.

$$\max_{W} \sum_{k=0}^{K-1} \frac{(2\pi f_k)^2 |X(f_k)A(f_k)|^2}{\sigma_k^2} \quad \text{s.t. } |w_{m,n}| = 1. \tag{3.51}$$

Express the summation, and the numerator can be rewritten as follows

$$\begin{aligned}(2\pi f_k)^2|X(f_k)A(f_k)|^2 &= |X(f_k)|^2 Tr\left[\boldsymbol{E}(f_k)^T\boldsymbol{W}\right] \\ &= (2\pi f_k)^2|X(f_k)|^2\left(vec(\boldsymbol{E}(f_k))^T vec(\boldsymbol{W})\right)^2 \\ &= (2\pi f_k)^2|X(f_k)|^2\left(\boldsymbol{e}(f_k)^T\boldsymbol{w}\right)^2 \\ &= (2\pi f_k)^2|X(f_k)|^2\left(\boldsymbol{w}^H\boldsymbol{e}(f_k)^*\boldsymbol{e}(f_k)^T\boldsymbol{w}\right) \\ &= \boldsymbol{w}^H\boldsymbol{\Sigma}_{XAE}(f_k)\boldsymbol{w}.\end{aligned} \tag{3.52}$$

The denominator can be expressed as

$$\begin{aligned}\sigma_k^2 &= \sigma_n^2 Tr\left[\boldsymbol{W}\boldsymbol{e}_n(f_k)\boldsymbol{e}_n(f_k)^H\boldsymbol{W}^H\right] \\ &= \sigma_n^2\boldsymbol{w}^T\left[e_n(f_k)e_n(f_k)^H \otimes I_{M\times M}\right]\boldsymbol{w}^* \\ &= \boldsymbol{w}^H\boldsymbol{\Sigma}_{NE}(f_k)\boldsymbol{w}.\end{aligned} \tag{3.53}$$

In which,

$$e_n(f_k) = \left[1, \ldots, e^{j2\pi f_k(N-1)\Delta T}\right]^T, \tag{3.54}$$

$$e_m(f_k) = \left[1, \ldots, e^{j2\pi(f_c+f_k)(M-1)\Delta\tau}\right]^T. \tag{3.55}$$

Thus, after transformation, the original form of summation can be transformed into the following form.

$$\sum_{k=0}^{K-1} \frac{(2\pi f_k)^2 |X(f_k)A(f_k)|^2}{\sigma_k^2} = \sum_{k=0}^{K-1} \frac{\boldsymbol{w}^H \boldsymbol{\Sigma}_{XAE}(f_k)\boldsymbol{w}}{\boldsymbol{w}^H \boldsymbol{\Sigma}_{NE}(f_k)\boldsymbol{w}}. \tag{3.56}$$

Obviously, this problem is an optimization problem for multiple generalized Rayleigh quotient summations, which is caused by the unique structure of the TDL array. In the TDL structure, the delay line and phase shifter behind each antenna together form a frequency-selective structure. When the frequency selection structures of multiple antennas are summed, the entire TDL matrix changes the SNR of the signals at different frequencies. Therefore, in the above summation, not only the signal term corresponding to the molecule is frequency-dependent, but also the noise term corresponding to the denominator is also frequency-dependent. Considering that the fractional summation problem is extremely difficult to handle, we amplify the noise term to simplify the problem, so there is

$$\sum_{k=0}^{K-1} \frac{\boldsymbol{w}^H \boldsymbol{\Sigma}_{XAE}(f_k)\boldsymbol{w}}{\boldsymbol{w}^H \boldsymbol{\Sigma}_{NE}(f_k)\boldsymbol{w}} \leq \sum_{k=0}^{K-1} \frac{\boldsymbol{w}^H \boldsymbol{\Sigma}_{XAE}(f_k)\boldsymbol{w}}{\boldsymbol{w}^H \boldsymbol{\Sigma}'_{NE}\boldsymbol{w}} = \frac{\boldsymbol{w}^H \boldsymbol{\Sigma}'_{XAE}\boldsymbol{w}}{\boldsymbol{w}^H \boldsymbol{\Sigma}'_{NE}\boldsymbol{w}}. \tag{3.57}$$

Our propose is to maximize the $\frac{\boldsymbol{w}^H \boldsymbol{\Sigma}'_{XAE}\boldsymbol{w}}{\boldsymbol{w}^H \boldsymbol{\Sigma}'_{NE}\boldsymbol{w}}$. However, the constant module optimization problem is a typical non-convex problem, it is difficult to directly obtain a closed-form solution, nor can it be solved directly by the convex optimization method. Therefore, we introduce a certain degree of relaxation in the processing process as shown in [41].

The constant modulus constraint problem is first relaxed into an inequality constraint problem in our chapter. The original optimization problem can be written as

$$\begin{aligned} &\max_{\mathrm{w}} \frac{\boldsymbol{w}^H \boldsymbol{\Sigma}'_{\boldsymbol{XAE}} w}{\boldsymbol{w}^H \boldsymbol{\Sigma}'_{\boldsymbol{NE}}\boldsymbol{w}} \\ &\text{s.t. } |\mathrm{w_i}| \leq \mathrm{P_i}/\boldsymbol{D}_{ii}. \end{aligned} \tag{3.58}$$

Then the global optimal solution to this problem can be written as

$$\mathbf{w} = \eta \mathbf{v}. \tag{3.59}$$

Among them, v is the main eigenvector of matrix $\boldsymbol{\Sigma}'^{-1}_{\mathbf{NE}}\boldsymbol{\Sigma}'_{\mathbf{XAE}}$, and η can be any value, for example, you can choose

$$\eta = \frac{1}{\left(\sqrt{\mathbf{D}_{kk}}\,|v_k|\right)/\sqrt{P_k}}. \tag{3.60}$$

v_k represents the **k**-th element in the vector **v**

$$\mathbf{k} = \arg\max_{1\leq i\leq r} \frac{\boldsymbol{D}_{ii}\,|v_i|^2}{P_i}. \tag{3.61}$$

For the equality constraints in the constant modulus optimization problem, the original optimization problem can be simplified as follows

$$\max_{W} \frac{\boldsymbol{w}^H \boldsymbol{\Sigma}'_{XAE} \boldsymbol{w}}{\boldsymbol{w}^H \boldsymbol{\Sigma}'_{NE} \boldsymbol{w}} \quad \text{s.t. } |w_{m,n}| = 1. \tag{3.62}$$

This problem is a typical generalized Rayleigh quotient problem if constant modulus constraints are not considered. Denote $\boldsymbol{u}$ as the primitive eigenvector of $\boldsymbol{\Sigma}'_{NE}{}^{-1/2} \boldsymbol{\Sigma}'_{XAE} \boldsymbol{\Sigma}'_{NE}{}^{-1/2}$,

$$\boldsymbol{w}^{opt} = \exp(j \cdot \angle \boldsymbol{w}). \tag{3.63}$$

3.5 Numerical Results and Analysis

In this section, we test the proposed algorithm by numerical simulation.

First, we verified the feasibility delay estimation under half-wave array conditions. The performances of the near CRLB estimation method are evaluated and compared with CRLB. Then, we verified that the TDL structure proposed in this chapter can still work better than the traditional array. Similarly, the result is compared with the optimal pilot and the pseudo-random pilot. At last, the simulation shows that the optimal TDL matrix outperforms the DFT matrix, and the performance increases linearly with the number of antennas.

3.5.1 Estimation Variance Versus CRLB

Figure 3.1 shows the comparison of the estimation method presented in this chapter with CRLB, in which the number of antennas of the antenna array is 16, 32, and 64, respectively. It can be found from the observation that in the case of a low SNR, the estimated variance given in this chapter is inferior to CRLB; however, as the SNR rises to the range of -15 to -5 dB, the performance of the proposed method is close to CRLB. The reason is that as the SNR increases, the 'arctan' function gradually shows a linearization trend. The conditions of the first-order Taylor expansion in the analysis gradually mature, and the phase estimation approximates the Gaussian distribution, which is consistent with the derivation process. As can be seen from the figure, as the number of antennas increases, the accuracy of delay estimation also increases.

What needs to be emphasized is that the received signal is under a relatively high SNR, because the core method proposed in this chapter uses the first-order approximation. Therefore, we specifically analyze the impact of the received signal on the delay estimation when the SNR of the received signal is low. For this case,

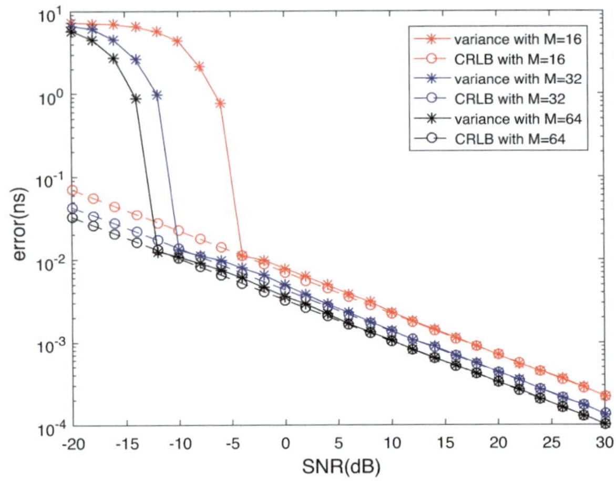

Fig. 3.1 Estimation variance versus CRLB under different numbers of antenna

the phase estimation method in this manuscript will indeed deteriorate. It can be found from Fig. 3.1 that for the case of low SNR, there is a large gap between the delay estimation performance of the phase method and the CRLB. The core of this phenomenon is that in a low SNR environment, the phase estimation performance of a single sampling point can no longer be approximated by first-order linearization, and the approximation error is very large. However, we can still improve the estimation accuracy through some methods. First, for a low SNR environment, due to the use of a long pilot sequence, the traditional matched filtering method can be used to find the optimal estimation of the delay. At this time, the matched filtering process is realized through the FFT technology, which also searches for signal phase information on multiple frequency components. According to the analysis, the phase information on different frequency components of the signal has the law of linear change. The phase method provided in this article first calculates the phase of a single frequency point and then calculates the optimal delay estimate by the linear method. However, when the SNR is relatively low, the core of the traditional MLE method is to first combine the signal energies of multiple frequency points and then perform phase estimation. The essence of this estimation method makes use of the delay information contained in the signal phase, but by merging the signals, the SNR can be improved, and the problem of deterioration of the method in this chapter under low SNR can be improved.

Meanwhile, it can be seen from Fig. 3.1 that the accuracy of delay estimation increases after the number of antennas increases. For large-scale arrays, an increase in the number of antennas is intuitively manifested as an increase in the received signal power. According to the analysis of our manuscript, the CRLB of the delay estimation is positively correlated with the SNR. Therefore, with 64 antennas, more signal power can be obtained. In the optimal design proposed in this chapter, the delay estimation result under 64 antennas is the best; while the delay estimation results for 32 antennas and 16 antennas decrease in order.

3.5.2 *TDL Structure Versus Traditional Array*

Figure 3.2 shows the comparison of the TDL structure proposed in this chapter with the tradition. Compared with the traditional narrow-band array, the TDL array proposed in this chapter has obvious advantages in the case of the number of antennas being 16, 32, and 64. What is particularly noticeable is that as the size of the antenna array increases, the delay estimation performance of the conventional array not only does not rise but decreases, so it is completely impossible under the conditions of large antenna size and high SNR. Meet the needs of the pass. This is because as the size of the antenna array increases, different antennas correspond to different data symbols, and only by phase-shifting the beat signals, the inter-code interference of the received signals is caused, and the SNR of the received signals is reduced and deteriorated. Estimate performance. Therefore, the traditional narrowband array can not only be used for the advantages of large-scale antenna arrays, but will be limited by its size, and the communication rate must be reduced. The TDL array has successfully solved this problem, making full use of the advantages brought by the increase in the size of the antenna array, and gradually reducing the estimation error.

3.5.3 *Optimal Training Pilot*

Figure 3.3 shows the synchronization performance comparison between the optimal training sequence and the traditional pseudo-random sequence proposed in this chapter. When the number of antennas is 16, 32, and 64, the delay estimation accuracy of the optimal training sequence is much higher than that of the traditional pseudo-random sequence. This is because, in an actual communication system, the energy

Fig. 3.2 TDL structure versus traditional narrowband array

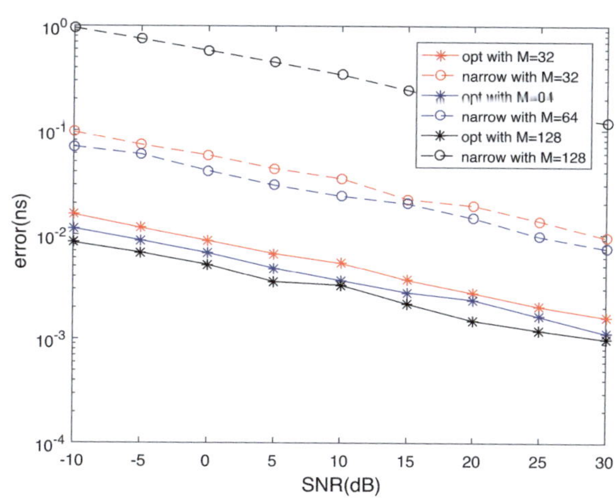

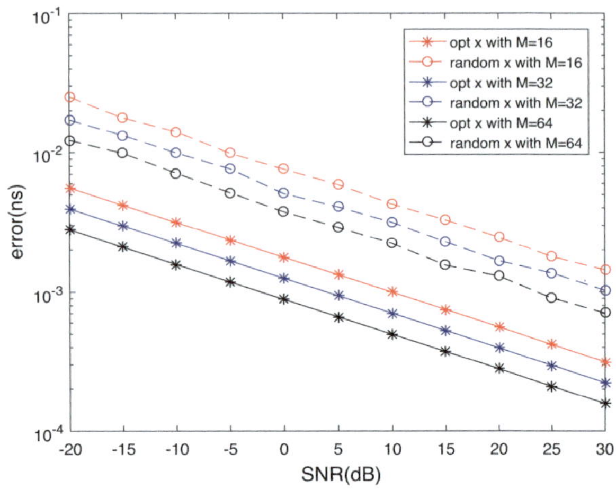

Fig. 3.3 Optimal training pilot versus pseudo-random pilot

of the signal is mainly concentrated near the baseband zero frequency, such as a typical sigma spectrum, and the high-frequency component is artificially suppressed. According to the discussion earlier in this chapter, the high-frequency signal is more helpful in improving the accuracy of the delay estimation. Therefore, in this regard, the conventional pseudo-random pilot does not guarantee that the delay estimation error is the smallest. The training sequence design method proposed in this chapter takes time delay estimation as the goal, rethinks the distribution of signal energy at each frequency point, introduces the weighting coefficient $(2\pi f_k)^2$, consciously increases the proportion of high-frequency signals, and then in the constant envelope. The delay estimation performance of the signal is improved under constraints.

3.5.4 Optiaml TDL Matrix

Figures 3.4 and 3.5 show the performance comparison between the optimal TDL array proposed in this chapter and the traditional DFT-based TDL matrix. Compared with the narrowband array structure, the DFT matrix-based TDL structure not only collects the signal energy of the center frequency point, but also collects the signal energy around the center frequency point, and thus the distortion due to excessive signal bandwidth. However, by comparison, it can be found that the optimal TDL matrix is superior to the DFT matrix in the case where the number of antennas is 16, 32, and 64, and in the case where the delayed series are sequentially 4, 6, and 8. Moreover, even with the TDL structure, the DFT matrix cannot solve the problem of reaching the asynchronous. Therefore, in the case where the number of antennas is relatively small, the DFT matrix can gradually improve the performance, but when the number of antennas is relatively large, the DFT matrix also has a performance

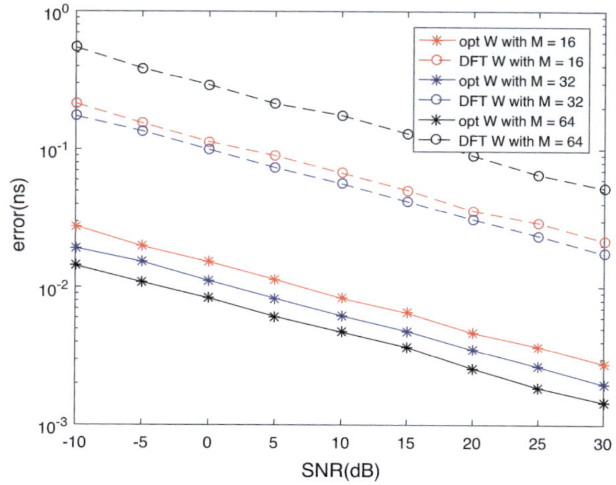

Fig. 3.4 Optiaml TDL matrix versus DFT matrix under different numbers of antennas

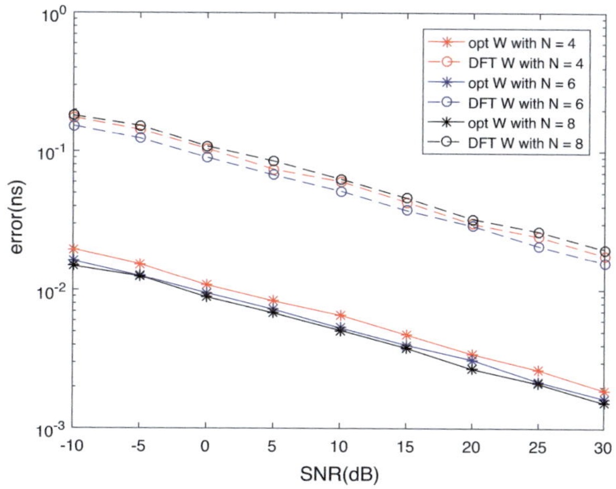

Fig. 3.5 Optiaml TDL matrix versus DFT matrix under different delay orders

deterioration problem. Moreover, the DFT matrix does not take full advantage of the advantages of the TDL architecture.

As can be seen from Fig. 3.4, for the TDL structure formed by the DFT matrix, the performance of 64 antennas is the best, and the performance of 16 antennas exceeds 32 antennas. This is because, for large-scale arrays, an increase in the number of antennas is intuitively manifested as an increase in the received signal power. According to the analysis of the article, the CRLB of the delay estimation is positively correlated with the SNR. Therefore, with 64 antennas, more signal power can be obtained. In the optimal design proposed in our article, the delay estimation result is the best; while the delay estimation results for 32 antennas and 16 antennas decrease in order. In contrast, for wide-band large-scale arrays, due to the increase in signal bandwidth, new influencing factors are introduced. For large-scale arrays, due

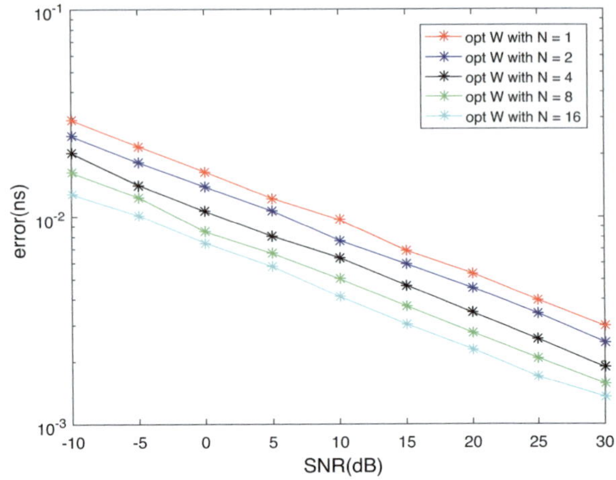

Fig. 3.6 Performance of optimal TDL matrix under different delay orders

to the sharp increase in the size of the array, the correlation of the received signals on each antenna is weakened, or even completely uncorrelated. Therefore, the use of a traditional DFT matrix for multi-antenna signal combination will suppress certain frequency components in the broadband signal and deteriorate the SNR of the received signal. In this way, the array gain no longer increases with the increase of the array size, and a larger-scale array destroys the received signal structure. Therefore, when the DFT array structure is adopted, it appears that the small-scale array has higher delay estimation accuracy.

Figure 3.6 shows the performance comparison of the optimal TDL matrix proposed in this chapter under different delay levels. By comparison, the performance of the optimal TDL array is increasing with the increase of the delayed series. The performance improvement comes from two aspects. On the one hand, as the number of delay stages increases, the constant-mode TDL structure can approximate an ideal delay to achieve signal alignment on each antenna; on the other hand, as the number of delay stages increases The signal of a longer time enters the TDL structure, and the equivalent SNR is also improved, thereby improving the accuracy of the delay estimation.

3.6 Conclusion

In this chapter, we have proposed a new constant modulus TDL structure for the THz large-scale antenna array synchronization problem. Based on the proposed method, we have derived the CRLB for synchronous delay estimation and given a time delay estimation method that can gradually reach the CRLB. By minimizing the CRLB, the design method of the optimal training sequence and the optimization method of the optimal TDL matrix are given. Simulation results have shown that with the improvement of SNR, the proposed estimation method can quickly converge to reach the CRLB. Compared with the method of using the training pilot of the traditional pseudo-random sequence, the optimal pilot designed in this chapter can achieve higher delay estimation accuracy. As the antenna size increases, the performance of the proposed TDL structure significantly exceeds the existing narrowband array structure. Furthermore, the performance of the optimal TDL matrix is also superior to that of the traditional DFT matrix-based structure.

References

1. Sampath H, Stoica P, Paulraj A (2001) Generalized linear precoder and decoder design for MIMO channels using the weighted MMSE criterion. IEEE Trans Commun 49(12):2198–2206
2. Palomar DP, Cioffi JM, Lagunas MA (2003) Joint Tx-Rx beamforming design for multicarrier MIMO channels: a unified framework for convex optimization. IEEE Trans Signal Process 51(9):2381–2401
3. Li L, Wang D, Niu X et al (2018) MmWave communications for 5G: implementation challenges and advances. Sci China Inf Sci 61(2):021301
4. Song H, Nagatsuma T (2011) Present and future of terahertz communications. IEEE Trans Terahertz Sci Technol 1(1):256–263
5. Kleine-Ostmann T, Nagatsuma T (2011) A review on terahertz communications research. J Infrared Millim Terahertz Waves 32:143–171
6. Han C, Akyildiz IF, Gerstacker WH (2017) Timing acquisition and error analysis for pulse-based terahertz band wireless systems. IEEE Trans Veh Technol 66(11):10102–10113
7. Liu Q, Wang Y, Fathy AE (2012) A compact integrated 100 Gs/s sampling module for UWB see through wall radar with fast refresh rate for dynamic real time imaging. In: IEEE radio and wireless symposium. IEEE, pp 59–62
8. Sarantos CH, Dagli N (2010) A photonic analog-to-digital converter based on an unbalanced Mach-Zehnder quantizer. Opt Express 18(14):14598–14603
9. Han C, Bicen AO, Akyildiz IF (2016) Multi-wideband waveform design for distance-adaptive wireless communications in the terahertz band. IEEE Trans Signal Process 64(4):910–922
10. Wang B, Gao F, Jin S et al (2018) Spatial- and frequency-wideband effects in millimeter-wave massive MIMO systems. IEEE Trans Signal Process 66(13):3393–3406
11. Jian M, Gao F, Tian Z et al (2019) Angle-domain aided UL/DL channel estimation for wideband mmWave massive MIMO systems with beam squint. IEEE Trans Wirel Commun 18(7):3515–3527
12. Xin R, Ni Z, Wu S et al (2019) Low-complexity joint channel estimation and symbol detection for OFDMA systems. China Commun 16(7):49–60
13. Gul MMU, Ma X, Lee S (2015) Timing and frequency synchronization for OFDM downlink transmissions using Zadoff-Chu sequences. IEEE Trans Wirel Commun 14(3):1716–1729
14. Amor SB, Affes S, Bellili F et al (2019) Multi-node ML time and frequency synchronization for distributed MIMO-relay beamforming over time-varying flat-fading channels. IEEE Trans Commun 67(4):2702–2715
15. Zhang R, Xia W, Yan F et al (2019) A single-site positioning method based on ToA and DoA estimation using virtual stations in NLOS environment. China Commun 16(2):146–159
16. Feng R, Liu Y, Huang J et al (2018) Wireless channel parameter estimation algorithms: recent advances and future challenges. China Commun 15(5):211–228
17. Shen X, Liao Y, Dai X et al (2018) Joint channel estimation and decoding design for 5G-enabled V2V channel. China Commun 15(7):39–46
18. Cheng F, Gui G, Zhao N et al (2019) UAV-relaying-assisted secure transmission with caching. IEEE Trans Commun 67(5):3140–3153

19. Zhao N, Lu W, Sheng M et al (2019) UAV-assisted emergency networks in disasters. IEEE Wirel Commun 26(1):45–51
20. Zhao N, Pang X, Li Z et al (2019) Joint trajectory and precoding optimization for UAV-assisted NOMA networks. IEEE Trans Commun 67(5):3723–3735
21. Gao Y, Xu J, Jia X (2015) Joint number and DoA estimation via the eigen-beam mCapon method for closely spaced sources. Sci China Inf Sci 58(12):217–219
22. Xiong X, Deng Z, Qi W et al (2019) High-precision angle estimation based on phase ambiguity resolution for high resolution radars. Sci China Inf Sci 62(4):53–55
23. Rong B, Zhang Z, Zhao X et al (2019) Robust superimposed training designs for MIMO relaying systems under general power constraints. IEEE Access 7:80404–80420
24. Ge Y, Zhang W, Gao F et al (2019) Angle-domain approach for parameter estimation in high-mobility OFDM with fully/partly calibrated massive ULA. IEEE Trans Wirel Commun 18(1):591–607
25. Stein Ioushua S, Yair O, Cohen D et al (2017) CaSCADE: compressed carrier and DoA estimation. IEEE Trans Signal Process 65(10):2645–2658
26. Gupta P, Agrawal M (2019) Design and analysis of the sparse array for DoA estimation of noncircular signals. IEEE Trans Signal Process 67(2):460–473
27. Lin B, Liu J, Xie M et al (2015) Direction-of-arrival tracking via low-rank plus sparse matrix decomposition. IEEE Antennas Wirel Propag Lett 14:1302–1305
28. Qin G, Amin MG, Zhang YD (2019) DoA estimation exploiting sparse array motions. IEEE Trans Signal Process 67(11):3013–3027
29. Guo M, Zhang YD, Chen T (2018) DoA estimation using compressed sparse array. IEEE Trans Signal Process 66(15):4133–4146
30. Yang X, Sun Y, Zeng T et al (2016) Fast STAP method based on PAST with sparse constraint for airborne phased array radar. IEEE Trans Signal Process 64(17):4550–4561
31. Bar-Ilan O, Eldar YC (2014) Sub-Nyquist radar via doppler focusing. IEEE Trans Signal Process 62(7):1796–1811
32. Cohen D, Cohen D, Eldar YC et al (2018) SUMMeR: Sub-Nyquist MIMO radar. IEEE Trans Signal Process 66(16):4315–4330
33. Aberman K, Eldar YC (2017) Sub-Nyquist SAR via Fourier domain range-doppler processing. IEEE Trans Geosci Remote Sens 55(11):6228–6244
34. Zheng L, Wang X (2017) Super-resolution delay-doppler estimation for OFDM passive radar. IEEE Trans Signal Process 65(9):2197–2210
35. Donoho DL, Maleki A, Montanari A (2009) Message passing algorithms for compressed sensing. Proc Natl Acad Sci 106(45):18914–18919
36. Rangan S (2011) Generalized approximate message passing for estimation with random linear mixing. In: IEEE international symposium on information theory proceedings, pp 2168–2172 (ISIT)
37. Al-Shoukairi M, Schniter P, Rao BD (2018) A GAMP-based low complexity sparse Bayesian learning algorithm. IEEE Trans Signal Process 66(2):294–308
38. Vila J, Schniter P (2011) Expectation-maximization Bernoulli-Gaussian approximate message passing. In: Asilomar conference on signals, systems, and computers, pp 799–803
39. Guo C, Nelson JDB (2015) Direction of arrival estimation via approximate message passing. In: IET conference, pp 1–6
40. Wei C, Liu H, Zhang Z et al (2017) Approximate message passing-based joint user activity and data detection for NOMA. IEEE Commun Lett 21(3):640–643
41. Havary-Nassab V, Shahbazpanahi S, Grami A et al (2008) Distributed beamforming for relay networks based on second-order statistics of the channel state information. IEEE Trans Signal Process 56(9):4306–4316

Chapter 4
Accurate AoA Tracking Method in MIMO Communication Systems

Yunzhu Xian, Chen Zheng, and Xuhui Ding

In this chapter, we investigate the angle tracking problem in a THz-based large-scale antenna array using a hybrid structure. Section 4.1 introduces the motivation of studying accurate AoA tracking method in MIMO communication systems. Section 4.2 describes the system structure and gives out the received signal model. In Sect. 4.3, we present the initial channel estimation method by using a random phase matrix. Furthermore, Sect. 4.4 describes the angle tracking method. Section 4.5 evaluates the performance of the proposed algorithm through numerical simulations. Finally, conclusions are drawn in Sect. 4.6.

4.1 Introduction

To meet the growing demand for high data rate communication, MIMO technology has been widely regarded as one of the most important and fundamental technologies in future wireless communication networks [1–4]. To obtain higher degrees of freedom, the scale of MIMO systems must be continuously increased. However, considering the massive MIMO technology, which usually uses hundreds or thousands of antenna elements, will be widely used in the 5G and beyond systems, the above extensive technology development route will soon encounter bottlenecks. Another

Y. Xian · C. Zheng · X. Ding (✉)
Beijing Institute of Technology, Beijing, China
e-mail: dingxuhui@bit.edu.cn

Y. Xian
e-mail: xianyunzhu@bit.edu.cn

C. Zheng
e-mail: mz3721@126.com

X. Ding et al. (eds.), *Performance Analysis and Improvement in MIMO Communication Systems*, https://doi.org/10.1007/978-981-96-5951-7_4

key technology of the 5G system, namely the THz technology, has the characteristics of large bandwidth and short wavelength. Since this technology significantly reduces the physical size of the antenna compared with microwave technology, it can bring new opportunities for further development and wide application of MIMO technology. In theory, the maximum number of antenna elements supported by THz technology is dozens of times that of traditional microwave arrays. In particular, due to the benefits such as high band, weak signal scattering, and angular sparseness, THz technology is capable of supporting multiple spatially separate users at the same time, making it suitable for combining with massive MIMO systems.

Nevertheless, THz technology still faces many challenges. For example, limited by the art of the state, the power of the THz amplifier is relatively small, and the THz signal is easily attenuated rapidly during transmission. To combat the above issue, channel estimation with high accuracy is highly demanded [5–7], because even a small angular change may induce a serious performance degradation [8, 9]. Traditional channel estimation and tracking methods highly rely on scanning technology to search the entire angle space, which becomes unsuitable for THz communication systems. In particular, in high-speed communication systems, frequent beam scanning is usually required to perform, causing a huge pilot overhead and thus reducing seriously the system's resource utilization. To address this challenge, high-performance channel estimation for massive MIMO has been widely investigated [9, 10].

Considering a large number of antennas will be comprised in a massive MIMO system, a full-digital structure may significantly increase both the cost and design complexity of the system. To simplify the system design, a hybrid structure is usually employed, in which the number of RF chains can be reduced by using a team of shifters and combiners [11, 12]. Furthermore, an analog matrix-based on a DFT matrix is usually employed by the hybrid structure [13–16]. It should be emphasized that reducing the number of RF chains does not always bring performance gains, because the reduction in the number of RF chains is accompanied by a reduction in the number of observations of the signal. In addition, because the beamwidth of a large-scale antenna system is extremely narrow, a large spatial range must be searched during the AoA estimation, which cannot be achieved with a limited RF chain. Even if we adopt some advanced technique (such as CS framework [17–19], the simulation structure based on the DFT matrix can hardly bear the burden of the observation matrix. Therefore, it is necessary to investigate a new AoA tracking method for hybrid structures [20–22].

The large-scale antenna system will provide corresponding services for different types of users. Because these users have different location information, their requirements for angle tracking are also different [23, 24]. For instance, for pedestrian users, using a sharper and narrower beam can achieve good user tracking; in contrast, for high-speed moving targets such as passengers and drones, a wider beam is required to track and provide services for these highly dynamic users. In addition, considering the time-varying characteristics of user location and channel information, AoA estimation must be completed in a short time to obtain the ideal tracking effect [25–27]. The continuous movement of the user inevitably leads to the continuous change of the beam angle. For example, for a user who is 2 m away from the BS and moves at a speed of 1 m/s, its displacement within 1 s will cause a 30° beam angle change. For a system with a beamwidth of about 0.5°, the above angular change requires the BS to perform AoA estimation at least 60 or even 100 times per second to guarantee that the user is under-tracking of the beam.

In essence, the AoA estimation has a typical nonlinear characteristic [28–31]. According to the mathematical expression of the received signal, the AoA mainly affects the phase of the received signal, whereas the AoA itself is in a sine function in digital space frequency. Therefore, the large-scale array has serious nonlinear characteristics for the AoA estimation problem. To address this kind of non-linear problem, a locally linearized extended Kalman filter (EKF) or unscented Kalman filter model is usually employed in traditional AoA tracking techniques [19, 28, 31]. For large-scale systems, on the other hand, new tracking methods are highly demanded due to the extremely sharp AoA and the risk of destroying locally linear approximations due to a mall angular change.

4.2 System Structure and Signal Model

Consider a BS equipped with a large-scale array THz communication system that is capable of serving K users simultaneously. To reduce the number of RF chains, a hybrid structure can be employed, as shown in Fig. 4.1, in which the received signals at M antennas are combined into N RF signals by using an analog matrix. Due to the users' mobility, the channel parameters of each user must be estimated and tracked by the BS to ensure a reliable CSI is available at the BS during the communication process.

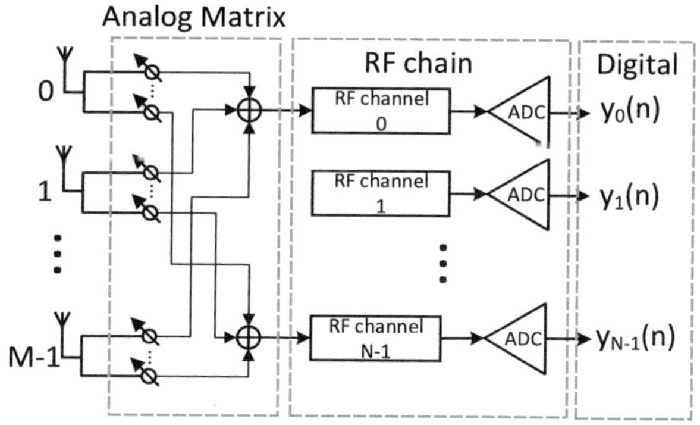

Fig. 4.1 The BS equipped with a hybrid structure uses a large-scale half-wave array with M antennas to enable M received signals at these antennas to be combined into N signals by using a $N \times M$ analog matrix and to be processed by N RF chains

4.2.1 Parameter Update Model

In the THz band, the signal transmission mainly depends on the LoS path, because both the reflection and scattering effects of the signal are extremely weak. Thanks to the short duration (i.e., on the order of ms) of THz communication, the changes in the signal strength are so small as to be ignored. Therefore, the motion process of the parameters in the time-varying channel can be modeled as

$$\boldsymbol{\theta}(n) = \boldsymbol{\theta}(n-1) + \boldsymbol{w}(n) + \boldsymbol{u}(n), \tag{4.1}$$

where $\boldsymbol{\theta}(n)$ corresponds to the angle of the channel at time n, $\boldsymbol{w}(n)$ represents the change of the angle caused by the user's movement in two adjacent moments, $\boldsymbol{u}(n)$ denotes the error of the angle at time n. In practice, it is difficult for us to directly obtain the exact value of $\boldsymbol{w}(n)$, that's why we rewrite the model by taking $\boldsymbol{w}(n)$ as a state variable. To estimate $\boldsymbol{w}(n)$, we introduce another error term $\boldsymbol{v}(n)$, relying on which we can estimate $\boldsymbol{w}(n)$ by using a tracking method.

$$\begin{bmatrix} \boldsymbol{\theta}(n) \\ \boldsymbol{w}(n) \end{bmatrix} = \begin{bmatrix} \boldsymbol{I} & \boldsymbol{I} \\ \boldsymbol{0} & \boldsymbol{I} \end{bmatrix} \begin{bmatrix} \boldsymbol{\theta}(n-1) \\ \boldsymbol{w}(n-1) \end{bmatrix} + \begin{bmatrix} \boldsymbol{u}(n) \\ \boldsymbol{v}(n) \end{bmatrix}. \tag{4.2}$$

4.2.2 Initial Pilot Model

Consider a BS of a large-scale array system with M antennas, which can serve K single-antenna users simultaneously. The model of the received signal can be expressed as

$$\boldsymbol{Y} = \boldsymbol{G}_A \boldsymbol{H} \boldsymbol{P} + \boldsymbol{G}_A \boldsymbol{W}, \tag{4.3}$$

where $\boldsymbol{G}_A \in C^{N \times M}$ represents the analog matrix, $\boldsymbol{H} = \left[\boldsymbol{h}_0, \boldsymbol{h}_1, \ldots, \boldsymbol{h}_{K-1}\right] \in C^{M \times K}$ denotes the channel matrix of K users, $\boldsymbol{h}_k$ stands for channel vector of user k, and $\boldsymbol{P} = \left[\boldsymbol{p}_0, \boldsymbol{p}_1, \ldots, \boldsymbol{p}_{K-1}\right] \in C^{K \times T_L}$ represents pilots of K users, each of which contains T_L symbols. Since the initial long pilots are orthogonal to each other, by using the matched filter, we get

$$\begin{aligned} \boldsymbol{r}_k = \boldsymbol{Y} \boldsymbol{p}_k^* &= \boldsymbol{G}_A \boldsymbol{h}_k \left[\boldsymbol{p}_k^T \boldsymbol{p}_k^*\right] + \boldsymbol{G}_A \boldsymbol{w}_k \\ &= T_L \sigma_k^2 \boldsymbol{G}_A \boldsymbol{h}_k + \boldsymbol{G}_A \boldsymbol{w}_k. \end{aligned} \tag{4.4}$$

To reduce the complexity of the proposed system, a digital-analog hybrid structure is usually employed. In the proposed model, $\boldsymbol{G}_A$ denotes a flat analog matrix with $N \ll M$. Therefore, the dimension of the received signal $\boldsymbol{r}_k$ is much smaller than that of the channel $\boldsymbol{h}_k$. As a result of the rank deficit, the LMMSE estimation cannot be implemented. However, thanks to channel sparsity in the angular domain, we can use the CS method to estimate the channel $\boldsymbol{h}_k$. First, consider the corresponding array manifold of the user k, as follows

$$\boldsymbol{h}_k = \left[e^{j2\pi f 0 \frac{d sin\theta_k}{c}} \cdots e^{j2\pi f (M-1) \frac{d sin\theta_k}{c}} \right]^T, \tag{4.5}$$

where θ_k denotes the angle of user k.

The problem of initial channel parameter estimation can thus be written as

$$\hat{\theta}_k = \min_{\theta_k} |\boldsymbol{r}_k - T_L \sigma_k^2 \boldsymbol{G_A} \boldsymbol{h}_k(\theta_k)|_F^2. \tag{4.6}$$

Since $\boldsymbol{h}_k$ is a complex sinusoidal signal, parameter estimation can be performed by using spectrum estimation. However, the number of RF chains is far less than that of antennas, so choosing the sub-matrix of the DFT as the analog matrix $\boldsymbol{G_A}$ is evidently improper. Since once $\boldsymbol{h}_k$ falls into the zero space of the $\boldsymbol{G_A}$, it will result in $\boldsymbol{G_A}\boldsymbol{h}_k = 0$, implying that it is impossible to estimate the angle at all.

To address this issue, we can use the beam scanning method by selecting the sub-matrix of the DFT matrix as the $\boldsymbol{G_A}$ matrix. However, due to the limited number of RF chains, multiple scans are still required. To reduce overhead, we propose a CS-based estimation method, as shown in Fig. 4.2. By using a random analog matrix, we can utilize the sparsity of angle space for estimating the channel with a limited number of RF chains at one time.

After the first-stage channel estimation, we can use the estimated parameters to construct both the analog matrix $\boldsymbol{G_A}$ and the digital matrix $\boldsymbol{G_D}$ that will be used in the communication process. In particular, $\boldsymbol{G_A}$ is used to both cover the users' motion and offer high beam gain, while $\boldsymbol{G_D}$ is used to recover the signal accurately.

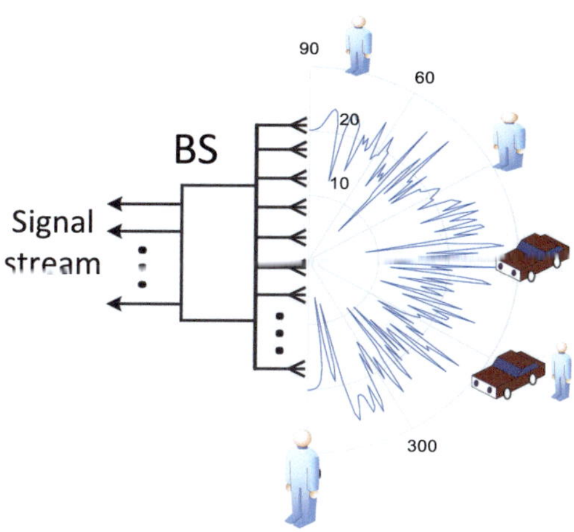

Fig. 4.2 Random analog matrix-based angle estimation in the initial stage. To avoid the disadvantage of using DFT beams, random analog matrix can be employed for constructing a CS structure

4.2.3 Tracking Pilot Model

In the tracking phase, short and nonorthogonal pilots are adopted. Short pilot symbols can bring many advantages. For example, by using the tracking method, short pilots can ensure a high estimation accuracy. At the same time, using short pilot symbols can save more resources for data transmission. For short pilots, the signal model at time n can be expressed as

$$\boldsymbol{Y}(n) = \boldsymbol{G}_A \boldsymbol{H}(n)\boldsymbol{q}(n) + \boldsymbol{G}_A \boldsymbol{w}(n), \tag{4.7}$$

where $\boldsymbol{q}(n) = [q_0(n), q_1(n), \ldots, q_{K-1}(n)]^T \in C^K$ represents K user's pilot, each of which contains one pilot symbol. The channel of user k can still be expressed as

$$\boldsymbol{h}_k(n) = \left[1 \cdots e^{j2\pi f(M-1)\frac{d sin\theta_k(n)}{c}}\right]^T. \tag{4.8}$$

The task of tracking is to estimate $\theta_k(n)$ based on the pilots as well as the past estimations in iterations $0 \sim (n-1)$. According to the current estimation, an updated digital matrix $\boldsymbol{G}_D(n)$ is designed for communication.

In some cases, the user's speed could be relatively slow (e.g., pedestrian), in which case we can use the DFT matrix to improve the beam gain. However, in some other cases, the users may move quickly (e.g., high-speed train), in which case the users may deviate out of the DFT beam since it is extremely narrow. Therefore, a wider beam must be employed, as shown in Fig. 4.3. Furthermore, the analog matrix $\boldsymbol{G}_A$ is not allowed to be changed frequently, but can only be designated at the initiating stage. Therefore, in the tracking process, only the digital matrix $\boldsymbol{G}_D$ (n) can be adjusted.

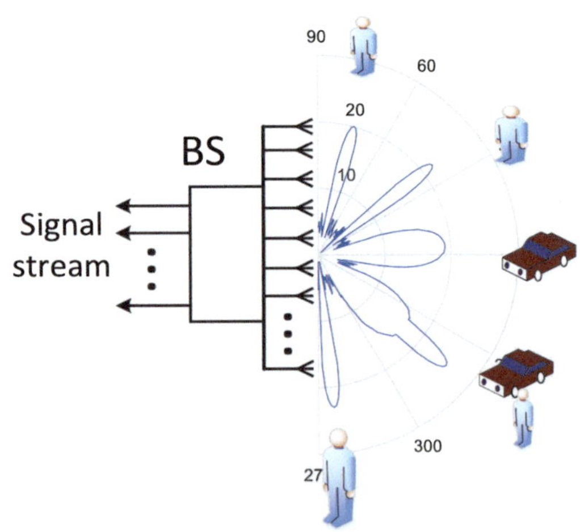

Fig. 4.3 Beam based angle estimation in the tracking stage. In the communication stage, to acquire a high gain and enhance signal power, both DFT beam and Wide beam can be used to point to the users. The DFT beam can be used for slowly moving users, while for fast moving users, the Wide beam is used

Above all, the two-stage angle tracking time domain frame structure is shown in Fig. 4.4. By adopting this timing model, we need to use a long pilot sequence once in the initial stage. In the tracking stage, the length of the pilot signal can be greatly reduced, thus more channel resources are allocated for communication data.

4.3 Initial Channel Estimation

In this section, we will introduce the proposed initial channel parameter estimation without prior knowledge through long pilots. Based on the results of channel parameter estimation, we can design an optimal analog matrix.

4.3.1 Random Matrix-Based Angle Estimation

Since the BS has no prior information on the angles at the initializing stage, the DFT based beam is not suitable now. By using a random analog matrix, a CS-based channel method can be proposed. In this chapter, we perform matched filtering to the received signal first. Relying on the matched filter, the interference between users can be eliminated while the power of the received signal can be enhanced. Consequently, the estimation accuracy can be substantially improved. For the channel parameter θ_k, the signal model can be further expressed as

$$\boldsymbol{r}_k = T_L \sigma_k^2 \boldsymbol{G}_A \boldsymbol{h}(\theta_k) + \boldsymbol{G}_A \boldsymbol{w}_k. \tag{4.9}$$

The entire angular space is discretized into several non-overlapping grids, where the signal only exists on the grid corresponding to the real angle. Therefore, the CS problem can be formulated. The long and orthogonal pilots lead to the fact that only one user is required to be estimated at one time, thus simplifying the problem solution. Furthermore, the random phase matrix $\boldsymbol{G}_A$ can be employed, avoiding the risk of missing signals that happens in the DFT matrix.

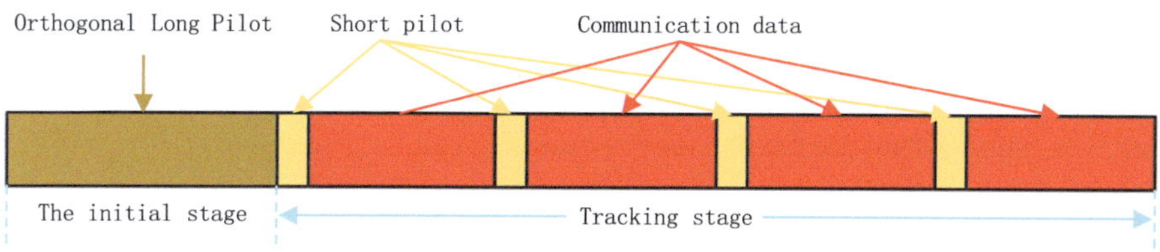

Fig. 4.4 A time domain frame structure of the tracking system

Constrained by the grids, a residual angle will remain. Therefore, based on the grid estimation, we can further search for θ near the exact estimation result $\theta_k^{'}$. Using the MLE algorithm, a more accurate estimation can be formulated as

$$\begin{aligned}\theta_k &= \max_{\theta\in[\theta_k^{'}-\Delta,\theta_k^{'}+\Delta]} \frac{1}{|T_L\sigma_k^2\boldsymbol{G_A h}(\theta)-r_k|^2}\\ &= \max_{\theta\in[\theta_k^{'}-\Delta,\theta_k^{'}+\Delta]} \frac{1}{A(\theta)}.\end{aligned} \tag{4.10}$$

Next, let us make such a definition

$$\begin{aligned}s(\theta) &= T_L\sigma_k^2\boldsymbol{G_A h}(\theta),\\ \frac{\partial s(\theta)}{\partial\theta} &= T_L\sigma_k^2\boldsymbol{G_A D h}(\theta),\end{aligned} \tag{4.11}$$

where

$$\boldsymbol{D} = \boldsymbol{diag}[0\cdots j\pi(M-1)]. \tag{4.12}$$

We can get

$$\begin{aligned}E[A(\theta)] &= E\left[|T_L\sigma_k^2\boldsymbol{G_A h}(\theta)-r_k|_F^2\right]\\ &\approx |\frac{\partial s(\theta)}{\partial\theta}(\theta-\theta_k)|_F^2 + Tr(\boldsymbol{G_A G_A}^H)\sigma_n^2\\ &\approx MN\sigma_n^2 + \frac{NM^3\pi^2}{3}(\theta-\theta_k)^2.\end{aligned} \tag{4.13}$$

Evidently, for $\theta=\theta_k$, $A(\theta)=MN\sigma_n^2$ which is approximately 0 at high-SNR regime. While for $\theta\neq\theta_k$, $A(\theta)$ grows rapidly. Since the pilots are long and orthogonal, the BS can achieve an estimation accuracy as high as CRLB, which can be taken as a benchmark.

4.3.2 Angular Range Analysis

To address the proposed CS problem, a random phases matrix is employed, in which the accuracy of the grid is limited by the array size. Therefore, the estimation error of the AoA can be denoted by $2\pi/M$. According to the motion model of the signal, the AoA range before the next channel estimation can be given by

$$\Delta\theta_k = w_k + u_k \approx w_k. \tag{4.14}$$

During the tracking procedure, the angular range can be expressed as

- When $\Delta\theta_k \geq 0$, in the current communication process (i.e., at time n), the possible AoA range is $[\theta_k(n)-2\pi/M, \theta_k(n)+2\pi/M+|\Delta\theta_k|]$.

- When $\Delta\theta_k < 0$, in the current communication process (i.e., at time n), the possible AoA range is $[\theta_k(n) - 2\pi/M - |\Delta\theta_k|, \theta_k(n) + 2\pi/M]$.

For convenience, we will denote this range by $B_k(n)$, representing the angular range at time n.

4.3.3 Analog Matrix Design Based on Initial Estimation

Even after we completed the initial parameter estimation, the analog matrix-based on the random phase distribution is still unsuitable for data transmission. Since such a matrix has a very low beam gain, it cannot take advantage of the large-scale array. Therefore, the analog matrix needs to be redesigned. A simple method for redesigning this matrix is to assign a certain RF link directly to a user, followed by taking the DFT matrix as a beam. This method is capable of both maximizing the signal power and suppressing interference. For user k, the responding beamformer can be expressed as

$$\boldsymbol{w}_k = \boldsymbol{h}_k(\theta)^*, \quad \boldsymbol{w}_k(m) = e^{-j2\pi f d m \frac{\sin\theta_k}{c}}. \tag{4.15}$$

Although the DFT beam can be employed for improving the signal's SNR, it is sensitive to the user's motion. To combat the user's motion, a wider beam is highly required. Therefore, we formulate the following problem of wider beam design

$$\boldsymbol{w}_k = \max_{\boldsymbol{w}} \min_{\theta\in[\theta_k-\Delta,\theta_k+\Delta]} |\boldsymbol{w}^H \boldsymbol{h}(\theta)|_F^2, \tag{4.16}$$

where $\boldsymbol{w}_k$ denotes the coefficient of the RF link corresponding to the k-th user. To satisfy the constraints of the analog matrix, the elements in $\boldsymbol{w}_k$ have a constant amplitude. For those wide beams pointing to different angles, we just make a rotation to the same wide beam with different angles. We first design a wide beam pointing at 0° as

$$\boldsymbol{w}_{wide} = \max_{\boldsymbol{w}} \min_{\theta\in[-\Delta,+\Delta]} |\boldsymbol{w}^H \boldsymbol{h}(\theta)|_F^2. \tag{4.17}$$

By performing a rotation

$$\boldsymbol{w}_k(m) = \boldsymbol{w}_{wide}(m) \times e^{-j2\pi f d \frac{m\sin\theta_k}{C}}, \tag{4.18}$$

we can redirect this beam to θ_k.

For optimizing the above problem, we can approximate it as follows

$$\max_{\boldsymbol{w}} \min_{\theta\in[-\Delta,+\Delta]} |\boldsymbol{w}^H \boldsymbol{h}(\theta)|_F^2 \approx \max_{\boldsymbol{w}} F_p(\boldsymbol{w}), \tag{4.19}$$

where $F_p(\boldsymbol{w}) = -\frac{1}{p}\log(\sum_{\theta\in[-\Delta,+\Delta]} e^{-p|\boldsymbol{w}^H \boldsymbol{h}(\theta)|_F^2})$. By implementing the gradient descent method, the above-mentioned optimization problem can be easily addressed.

4.4 Angle Tracking Method

The large-scale channel tracking studied in this chapter mainly comprises two issues:

- In practical scenarios, the angles may change fast. For high-speed moving users, the BS cannot directly use the narrow-beam DFT matrix, but can only seek a wider beam. For each user, only one RF chain can be assigned, making it hard to further improve the accuracy of angular estimation.
- The angle-observation model is a typical non-linear model, to which the traditional Kalman filter cannot be directly used. In the proposed angle tracking technique, observations (including $\phi_k = \sin\theta_k$ and $e^{j2\pi fm\frac{d\phi_k}{c}}$) are highly non-linear, in which the $\sin\theta_k$ can be approximated by local linear expansion. However, even a small angular change may cause a significant change $e^{j2\pi fm\frac{d\phi_k}{c}}$ in the large-scale array systems.

Therefore, to address the first problem, we first use the GAMP method to perform gridded angular search [32, 33]. After that, by using the modified expectation maximization (EM) algorithm, we can construct the linear observation of the signal [19, 34] for improving the accuracy of estimation.

4.4.1 GAMP-Based Angle Observation

First, we divide the entire space into J non-overlapping grids. According to the previous analysis, the user of interest should always be located inside the beam, thus reducing the search space. Considering that the THz beam is extremely narrow and the search duration is short, each user is allowed to occupy only one of the J discrete grids. When the system is serving U users simultaneously, whether there exist users on one beam for each grid (e.g., beam θ_j) obeys the Bernoulli distribution.

According to the model of the received signal in the tracking phase, we can rewrite it as

$$\begin{aligned}\boldsymbol{r} &= \sum_{k=1}^{K} \boldsymbol{G}_A \boldsymbol{h}(\theta_k) + \boldsymbol{G}_A \boldsymbol{n} \\ &= \boldsymbol{G}_A \boldsymbol{H} \boldsymbol{z} + \boldsymbol{G}_A \boldsymbol{n} \\ &= \boldsymbol{B}\boldsymbol{z} + \boldsymbol{\eta},\end{aligned} \tag{4.20}$$

where $\boldsymbol{z} \in C^J$ denotes the signal vector, and $\boldsymbol{H} \in C^{M\times J}$ is the extended channel matrix.

Since there are only U users, at most U elements in $\boldsymbol{z}$ can be 1, leaving all the other elements to be 0. In particular, $\boldsymbol{z}(i) = 1$ means that a user exists at the angle θ_i, and the corresponding channel vector is the i column in $\boldsymbol{H}$. In this way, we construct an extended observation model of the GAMP problem by defining the probability of user absent as λ.

Therefore, the distribution of $z(n)$ can be written as $p(z(n)) = \lambda_n \delta(z(n)) + (1 - \lambda_n)\delta(z(n) - 1)$, where $\lambda_n = (1 - \frac{1}{B_n})$ and B_n represents the beam width at θ_n. We use the central limit theorem to approximate the Gaussian distributions of both the intermediate variables $\boldsymbol{z}$ and $\boldsymbol{q}$. In this way, the conditional posterior distribution of $\boldsymbol{z}$ and $\boldsymbol{q}$ can be written as

$$\begin{aligned} P_{Z|S}(z|s; \hat{r}, \mu^r) &= \frac{P_Z(z) \times N(z; \hat{r}, \mu^r)}{\int_{z'} P_Z(z') \times N(\hat{z}; \hat{r}, \mu^r)} \\ P_{Q|S}(q|s; \hat{q}, \mu^q) &= N\left(q, \frac{s}{\sigma_n^2} + \frac{\hat{q}}{\mu^q}, \frac{\mu^q \sigma_n^2}{\mu^q + \sigma_n^2}\right). \end{aligned} \tag{4.21}$$

Due to the Gaussian approximation, we can easily calculate both the expectation and variance under the conditional posterior distribution of $\boldsymbol{z}$ as

$$\begin{aligned} E[Z|s; \hat{r}, \mu^r] &= P_{Z|Y}(1\, s; \hat{r}, \mu^r) \\ VAR[Z|s; \hat{r}, \mu^r] &= E[Z|s; \hat{r}, \mu^r] - E[Z|s; \hat{r}, \mu^r]^2. \end{aligned} \tag{4.22}$$

Meanwhile, since $P_{Q|S}(q|s; \hat{q}, \mu^q)$ is a product of two Gaussian distributions, we can easily calculate both the expectation and variance of $\boldsymbol{q}$ as

$$\begin{aligned} E_{Q|S}[q|s; \hat{q}, \mu^q] &= \frac{s}{\sigma_n^2} + \frac{\hat{q}}{\mu^q} \\ VAR_{Q|S}[q|s; \hat{q}, \mu^q] &= \frac{\mu^q \sigma_n^2}{\mu^q + \sigma_n^2}. \end{aligned} \tag{4.23}$$

Furthermore, we define new functions as

$$\begin{aligned} g_{in}(\hat{r}, \mu^r) &= E[Z|s; \hat{r}, \mu^r] \\ &= \left[1 + \frac{\lambda}{1-\lambda} \times \frac{N(0; \hat{r}, \mu^r)}{N(1; \hat{r}, \mu^r)}\right]^{-1} \\ g'_{in}(\hat{r}, \mu^r) &= \frac{1}{\mu^r} VAR[Z|s; \hat{r}, \mu^r] \\ &= \frac{1}{\mu^q}[g_{in}(\hat{r}, \mu^r) - g_{in}(\hat{r}, \mu^r)^2] \\ g_{out}(s, \hat{q}, \mu^q) &= \frac{1}{\mu^q}(E_{Q|S}[q|s; \hat{q}, \mu^q] - \hat{q}) \\ g'_{out}(s, \hat{q}, \mu^q) &= \frac{1}{\mu^q}\left(\frac{VAR_{Q|S}[q|s; \hat{q}, \mu^q]}{\mu^q} - 1\right). \end{aligned} \tag{4.24}$$

The algorithm can thus be described as Algorithm 4.1.

For z that needs to be estimated, the GAMP algorithm produces a list of estimates $z_n(t)$ and its corresponding variance $\mu_n^z(t)$.

Algorithm 4.1 GAMP-based angle observation algorithm

for each $n \in [0, N-1]$ **do**
  initialize $\hat{z}_n(1) = E(z) = (1-\lambda)$;
  initialize $\mu_n^z(1) = VAR(z) = (1-\lambda)\lambda$;
end for
for each $m \in [0, M-1]$ **do**
  initialize $\hat{u}_m(0) = 0$;
end for
while $t \le Iter_{max}$ **do**
  for each $m \in [0, M-1]$ **do**
    $\mu_m^p(t) = \sum_{n=0}^{N-1} |B_{m,n}|^2 \mu_n^z(t)$;
    $\hat{p}_m(t) = \sum_{n=0}^{N-1} B_{m,n}\hat{z}_N(t) - \mu_m^p(t)\hat{u}_m(t-1)$;
    $\hat{u}_m(t) = g_{out}(s_m, \hat{p}_m(t), \mu_m^p(t))$;
    $\mu_m^u(t) = -g_{out}^{'}(s_m, \hat{p}_m(t), \mu_m^p(t))$;
  end for
  for each $n \in [0, N-1]$ **do**
    $\mu_n^r(t) = (\sum_{m=0}^{M-1} |B_{m,n}|^2 \mu_m^u(t))^{-1}$;
    $\hat{r}_n(t) = \hat{z}_n(t) + \mu_n^r(t) \sum_{m=0}^{M-1} B_{m,n}^* \hat{u}_m(t)$;
    $\mu_n^z(t+1) = \mu_n^r g_{in}^{'}(\hat{r}_n(t), \mu_n^r(t))$;
    $\hat{z}_n(t+1) = g_{in}(\hat{r}_n(t), \mu_n^r(t))$;
  end for
end while

4.4.2 EM Based Instantaneous Angle Observation

After performing the GAMP method, we construct the linear Gaussian observation model of angle. For the large-scale arrays, the SNR gain could be very high, making the MLE estimation of ϕ_k progressive reaching CRLB. Thus, by using the modified EM method, the estimation follows the Gaussian distribution can be given by

$$\begin{aligned} \hat{\phi}_k &= \phi_k + n_{\phi_k} \\ n_{\phi_k} &\sim N(0, CRLB(\phi_k)). \end{aligned} \tag{4.25}$$

According to $\phi_k = sin\theta_k$, we get

$$\begin{aligned} \psi_k &= arcsin(\hat{\phi}_k) = \theta_k + n_{\theta_k}, \\ n_{\theta_k} &\sim N(0, CRLB(\theta_k)). \end{aligned} \tag{4.26}$$

Therefore, we could construct a linear observation model for the formulated problem. Furthermore, by using the conditions of MLE and high SNR of large-scale arrays, the signal observation model is guaranteed to have a Gaussian distribution. Intuitively speaking, ϕ_k can be obtained by using the grid-search method. However, in multi-user scenarios, employing the method of joint search would cause a dimensional disaster. In addition, in a high-SNR regime, the accuracy of the grid is at least one order of magnitude higher than that of CRLB, which induces a sharp increase in computational cost in order to approach CRLB.

In the following, we use the EM algorithm to construct a linear observation model for the angle in a high-SNR regime.

4.4.2.1 Likelihood Function Derivation

The received signal model in the tracking phase can be expressed as

$$\begin{aligned} \boldsymbol{y}(n) &= \boldsymbol{G}_A \boldsymbol{H}(n)\boldsymbol{q} + \boldsymbol{G}_A \boldsymbol{w}(n), \\ \boldsymbol{q} &= [q_1, q_2, \ldots, q_K]^T \in C^K. \end{aligned} \tag{4.27}$$

Since we can split the received signal into a superposition of K user signals, it can be further defined as

$$\begin{aligned} \boldsymbol{x}_i(n) &= \boldsymbol{G}_A \boldsymbol{h}_i(n) q_i + \boldsymbol{G}_A \boldsymbol{w}_i(n) \\ &= \boldsymbol{s}_i(n) + \boldsymbol{\eta}_i(n), \end{aligned} \tag{4.28}$$

where $\boldsymbol{s}_i(n) = \boldsymbol{G}_A \boldsymbol{h}_i(n) q_i$ and $\boldsymbol{\eta}_i(n) = \boldsymbol{G}_A \boldsymbol{w}_i(n)$, implying that at time n, the i-th user sends the pilot q_i through the corresponding channel $\boldsymbol{h}_i(n)$, following which the response $\boldsymbol{x}_i(n)$ is generated. After that, we get $\boldsymbol{y}(n) = \sum_{i=1}^{K} \boldsymbol{x}_i(n)$. Furthermore, we define

$$\boldsymbol{x}(n) = \left[\boldsymbol{x}_1{}^T, \boldsymbol{x}_2{}^T, \ldots, \boldsymbol{x}_K{}^T\right]. \tag{4.29}$$

The likelihood function can thus be written as

$$\begin{aligned} \ln p_X(\boldsymbol{x}|\theta) &= C + \sum_{i=1}^{K} \left[-\frac{1}{2} (\boldsymbol{x}_i - \boldsymbol{s}_i)^H \boldsymbol{\Sigma}_i^{-1} (\boldsymbol{x}_i - \boldsymbol{s}_i) \right] \\ &= C - \frac{1}{2} \left[\boldsymbol{x}^H \boldsymbol{\Sigma}^{-1} \boldsymbol{x} - 2Re(\boldsymbol{s}^H \boldsymbol{\Sigma}^{-1} \boldsymbol{x}) + \boldsymbol{s}^H \boldsymbol{\Sigma}^{-1} \boldsymbol{s} \right], \end{aligned} \tag{4.30}$$

where $\boldsymbol{\Sigma}_i$ represents the covariance matrix of $\boldsymbol{\eta}_i(n)$, $\boldsymbol{\Sigma}$ is a block diagonal matrix composed of $\boldsymbol{\Sigma}_i$. For a K-user scenario, according to the previous noise decomposition, the corresponding noise power of each user is only $1/K$ of the original noise power. According to the likelihood function, we can get the iterative form of the EM algorithm.

4.4.2.2 EM-Step

Following the likelihood function, we can get the conditional expectation as

$$U(\theta, \theta^t) = E\left[\ln p_X(\boldsymbol{x}|\theta)|y; \theta^t\right]. \tag{4.31}$$

Since we need to get the maximum value of $U(\theta, \theta^t)$, the items that are not related to θ can be omitted, implying that we need only to calculate

$$\begin{aligned} U^{'}(\theta, \theta^t) &= E\left[Re(\boldsymbol{s}^H \boldsymbol{\Sigma}^{-1} \boldsymbol{x}) - \frac{1}{2} \boldsymbol{s}^H \boldsymbol{\Sigma}^{-1} \boldsymbol{s} | y; \theta^t \right] \\ &= E\left[Re(\boldsymbol{s}^H \boldsymbol{\Sigma}^{-1} \boldsymbol{x}) | y; \theta^t \right] - \frac{1}{2} \boldsymbol{s}^H \boldsymbol{\Sigma}^{-1} \boldsymbol{s} \\ &= Re(\boldsymbol{s}^H \boldsymbol{\Sigma}^{-1} E\left[\boldsymbol{x} | y; \theta^t \right]) - \frac{1}{2} \boldsymbol{s}^H \boldsymbol{\Sigma}^{-1} \boldsymbol{s}. \end{aligned} \tag{4.32}$$

Thus, it is critical to compute $E\left[\boldsymbol{x}|y; \theta^t\right]$

$$\begin{aligned} E\left[\boldsymbol{x} \mid y; \theta^t\right] &= E(\boldsymbol{x}) + C_{xy} C_{yy}^{-1} [\boldsymbol{y} - E(\boldsymbol{y})] \\ &= \begin{bmatrix} s_1(\theta^t) \\ s_2(\theta^t) \\ \vdots \\ s_K(\theta^t) \end{bmatrix} + \frac{1}{N} \begin{bmatrix} \boldsymbol{y} - \sum_{i=1}^K s_i(\theta^t) \\ \boldsymbol{y} - \sum_{i=1}^K s_i(\theta^t) \\ \vdots \\ \boldsymbol{y} - \sum_{i=1}^K s_i(\theta^t) \end{bmatrix}. \end{aligned} \tag{4.33}$$

After that, we could rewrite (4.32) into (4.34), as follows

$$\begin{aligned} U'\left(\theta, \theta^t\right) = &\operatorname{Re}\left(\boldsymbol{s}(\theta)^H \boldsymbol{\Sigma}^{-1} \boldsymbol{s}\left(\theta^t\right)\right. \\ &\left. + \frac{1}{N} \boldsymbol{s}(\theta)^H \boldsymbol{\Sigma}^{-1} \begin{bmatrix} \boldsymbol{y} - \sum_{i=1}^K s_i(\theta^t) \\ \boldsymbol{y} - \sum_{i=1}^K s_i(\theta^t) \\ \vdots \\ \boldsymbol{y} - \sum_{i=1}^K s_i(\theta^t) \end{bmatrix}\right) \\ &- \frac{1}{2} \boldsymbol{s}(\theta)^H \boldsymbol{\Sigma}^{-1} \boldsymbol{s}(\theta). \end{aligned} \tag{4.34}$$

The maximization problem can be written as

$$\theta^{t+1} = \arg\max_{\theta} U^{'}(\theta, \theta^t). \tag{4.35}$$

Looking closely at the above problems, it is shown that the θ_i of the AoA of the i-th user can be estimated separately, without the need to estimate the K angles of arrival simultaneously. In this case, the problem can be further reduced to

$$\theta_i^{t+1} = \arg\max_{\theta} U^{''}(\theta, \theta^t), \tag{4.36}$$

where $U^{''}(\theta, \theta^t)$ can be expressed as

$$U''\left(\theta,\theta^{t}\right)=\mathrm{Re}\left(\boldsymbol{s}_i\left(\theta_i\right)^H\boldsymbol{\Sigma}_i^{-1}\boldsymbol{s}_i\left(\boldsymbol{\theta}^t\right)+\frac{1}{N}\boldsymbol{s}_i\left(\theta_i\right)^H\boldsymbol{\Sigma}_i^{-1}\left[\boldsymbol{y}-\sum_{i=1}^{K}\boldsymbol{s}_i\left(\boldsymbol{\theta}^t\right)\right]\right)$$
$$-\frac{1}{2}\boldsymbol{s}_i\left(\theta_i\right)^H\boldsymbol{\Sigma}_i^{-1}\boldsymbol{s}_i\left(\theta_i\right). \tag{4.37}$$

Furthermore, we can write the EM calculation as (4.38), as shown at the top of the next page. Through the EM algorithm, we can mitigate the huge computational complexity of direct joint search. However, we note that this optimization problem is still not easy to solve. Therefore, we use the beam characteristics of large-scale hybrid arrays to further simplify this problem.

$$\theta_i^{t+1}=\arg\max_{\theta_i}\mathrm{Re}\left(\boldsymbol{s}_i\left(\theta_i\right)^H\boldsymbol{\Sigma}_i^{-1}\boldsymbol{s}_i\left(\boldsymbol{\theta}^t\right)+\frac{1}{N}\boldsymbol{s}_i\left(\theta_i\right)^H\boldsymbol{\Sigma}_i^{-1}\left[\boldsymbol{y}-\sum_{i=1}^{K}\boldsymbol{s}_i\left(\boldsymbol{\theta}^t\right)\right]\right)$$
$$-\frac{1}{2}\boldsymbol{s}_i\left(\theta_i\right)^H\boldsymbol{\Sigma}_i^{-1}\boldsymbol{s}_i\left(\theta_i\right). \tag{4.38}$$

4.4.3 Simplified Large-Scale Array Problems

To simplify the solution of the maximization problem, we further examine the expression of $\boldsymbol{s}_i(\theta_i)$ as

$$\boldsymbol{s}_i(\theta_i)=\boldsymbol{G}_A\boldsymbol{h}_i(\theta_i)q_i. \tag{4.39}$$

For large-scale antenna systems, $\boldsymbol{s}_i(\theta_i)$ itself exhibits the following characteristics due to the use of a hybrid structure.

- In order to increase the gain of the antenna array, the system uses both narrow and wide beams to serve slow and fast users, respectively. The i-th element in the vector $\boldsymbol{s}_i(\theta_i)$ represents the i-th user's transmitted signal that passes through the corresponding beam combiner. Based on this, further utilizing both the beam's high gain and low sidelobe characteristics can reduce the complexity of the estimation algorithm.
- According to (4.38), in each iteration, the received signal $\boldsymbol{y}$ subtracts the last expected received signal $\sum_{i=1}^{K}\boldsymbol{s}_i(\boldsymbol{\theta}^t)$, following which it uses the remaining items for the current update. This inspires us that if the AoA of the other $(K-1)$ users is known, the estimate of the i-th user can be obtained directly from the rest of the received signal as

$$\hat{\theta}_i=\max_{\theta}|\boldsymbol{s}_i(\theta_i)^H\boldsymbol{\Sigma}_i^{-1}\left[\boldsymbol{y}-\sum_{i=1}^{K}\boldsymbol{s}_i(\boldsymbol{\theta}^t)\right]|^2. \tag{4.40}$$

The more accurate the estimate of the AoA for other $(K-1)$ users, the more accurate the AoA for users i. By employing a large-scale array, both the beam gain SNR can be high, thus enabling an estimation with a high enough accuracy.

- The hybrid structure has only one effective observation value for each user, to which the CS method is not applicable. However, it can be found that there is a correspondence between the arrival angle of the signal and the phase of the received signal, which can be estimated accordingly. Moreover, by taking the signal phase into account, we can avoid the deviation caused by the signal amplitude by eliminating the $\boldsymbol{s}_i(\theta_i)^H\boldsymbol{\Sigma}_i^{-1}\boldsymbol{s}_i(\theta_i)$ items. Therefore, the parameter estimation problem based on the beam phase can be rewritten as

$$\hat{\theta}_i = \min_{\theta_i} |\angle s_i(\theta_i) - \angle\left[\sum_{j\neq i}^{K} y_i - \boldsymbol{s}_j(\hat{\theta}_j)\right]|. \tag{4.41}$$

Therefore, by using the previous analysis, we can simplify the original optimization problem in the following manner. First, considering the fact that the received signal is the superposition of the signals of K users after K beam combiners, we can rewrite the form of the received signal as

$$\begin{aligned}\boldsymbol{y}(n) &= \boldsymbol{G_A H}(n)\boldsymbol{q} + \boldsymbol{G_A w}(n)\\ &= \begin{bmatrix} g_1^T h_1 & \cdots & g_1^T h_K \\ \vdots & \cdots & \vdots \\ g_K^T h_1 & \cdots & g_K^T h_K \end{bmatrix}\begin{bmatrix} q_1 \\ \vdots \\ q_K \end{bmatrix} + \boldsymbol{G_A w}(n),\end{aligned} \tag{4.42}$$

where the main diagonal element of the $\boldsymbol{G}_A\boldsymbol{H}(n)$ matrix implies that the user signal falls into its corresponding beam, in which case the signal amplitude is large. The off-diagonal elements of the matrix represent signals that other users fall into the current beam. Based on the above analysis, we can construct the following AoA search method:

- First, use the characteristics of narrow beams to make accurate AoA estimates for slow users served by narrow beams. Since the narrow beam gain is the highest, the attainable estimation accuracy is also the highest: $\hat{\boldsymbol{\theta}}_1 \sim p_1(\boldsymbol{\theta}_1|\boldsymbol{y})$.
- Based on the estimation results of the narrow beam, the signal components of the slow users are eliminated from the signal. Therefore, we perform an accurate search for the users in a wide beam: $\hat{\boldsymbol{\theta}}_2 \sim p_1(\boldsymbol{\theta}_2|\hat{\boldsymbol{\theta}}_1, \boldsymbol{y})$.
- Therefore, through the transmission of information between wide and narrow beams, we can get a preliminary estimate of the angle: $\hat{\boldsymbol{\theta}}^0 = \left[\hat{\boldsymbol{\theta}}_1, \hat{\boldsymbol{\theta}}_2\right]$.
- After completing the preliminary estimation, we can perform an iterative estimation: $\hat{\boldsymbol{\theta}}^{t+1} \sim p(\boldsymbol{\theta}|\hat{\boldsymbol{\theta}}^t, \boldsymbol{y})$.

In the following, we aim at estimation methods for both narrow- and wide-beam construction.

Narrow beam case
For the case of the DFT matrix, we can rewrite the received signal as

$$y(\Delta\phi) = \frac{\sin(\pi M \Delta\phi)}{\sin(\pi \Delta\phi)} e^{j\pi(M-1)\Delta\phi}. \tag{4.43}$$

It is shown the amplitude of the received signal decreases from the ideal value M to $\frac{\sin(\pi M \Delta\phi)}{\sin(\pi \Delta\phi)}$. There is an additional phase in the received signal, $e^{j\pi(M-1)\Delta\phi}$, which is directly related to the angular offset. Since $\Delta\phi$ is relatively small, this phase can be directly calculated. We can rewrite the signal observation model as

$$\boldsymbol{y} = \boldsymbol{A}(\Delta\phi)\boldsymbol{x}(\Delta\phi) + \boldsymbol{n}, \tag{4.44}$$

where

$$\begin{aligned} \boldsymbol{x}(\Delta\phi) &= \left[1 \cdots e^{j\pi(M-1)\Delta\phi_K}\right], \\ [\boldsymbol{A}(\Delta\phi)]_{i,j} &= \frac{\sin(\pi M(\phi_j - \phi_i + \Delta\phi_j))}{\sin(\pi(\phi_j - \phi_i + \Delta\phi_j))} e^{j\pi(M-1)(\phi_j - \phi_i)}. \end{aligned} \tag{4.45}$$

Obviously, since $\boldsymbol{A}(\Delta\phi)$ is related to the amount to be estimated $\Delta\phi$, we cannot get $\boldsymbol{A}(\Delta\phi)$ in advance. However, $\boldsymbol{A}(\Delta\phi)$ is approximately a diagonal matrix.

When $i = j$, we get

$$[\boldsymbol{A}(\Delta\phi)]_{i,i} = \frac{\sin(\pi M(\Delta\phi_j))}{\sin(\pi(\Delta\phi_j))} >> 1. \tag{4.46}$$

When $i \neq j$, we get

$$[\boldsymbol{A}(\Delta\phi)]_{i,j} << 1. \tag{4.47}$$

In practice, the spatial angles between users are very different, i.e., $|\phi_j - \phi_i| >> 1/M$ when $i \neq j$, making $[\boldsymbol{A}(\Delta\phi)]_{i,j} \approx 0$. Since the DFT matrix is used, the signals outside are suppressed to be $[\boldsymbol{A}(\Delta\phi)]_{i,j} \approx 0$, while the signal inside is enhanced as $[\boldsymbol{A}(\Delta\phi)]_{i,i} >> 1$. Thus, we can estimate the angle based on the phase of the received signal.

When $t = 0$, the estimation of $\Delta\phi$ is $\hat{\Delta\phi}^{(0)}$, leading to

$$\hat{\phi}_k^{(0)} = \phi_k + \hat{\Delta\phi}_k^{(0)}, \quad k = 1 \cdots K, \tag{4.48}$$

where $\hat{\Delta\phi}^{(0)} = 0$.

When $t > 0$, based on the previous estimation $\hat{\Delta\phi}^{(t-1)}$, we have

$$\hat{\phi}_k^{(t)} = \phi_k + \hat{\Delta\phi}_k^{(t)}, \quad k = 1, 2, \ldots, K, \tag{4.49}$$

where $\hat{\Delta\phi}^{(t)} = \frac{\angle \boldsymbol{A}(\hat{\Delta\phi}^{(t-1)})^{-1}\boldsymbol{y}}{\pi(M-1)}$.

Wide beam case

However, for a wide beam system, we cannot directly obtain the analytical relationship between the phase of the received signal and the angle. Fortunately, with the help of the GAMP method, the range of the user's angle can be reduced to inside the Rayleigh limit of the array in the high-SNR regime. The signal phase changes slowly and continuously as the angle changes inside the beam.

When the signal of k-th user falls into a certain beam, the noise as well as other users' signals are simplified to n. Furthermore, we can remove the slow user's signal from the received signal by means discussed in the narrow beam case. Because the large-scale array has a relatively high gain, the unknown interference and noise n are relatively small. We thus have

$$y(\Delta\phi_i) = \sum_{m=0}^{M-1} w_m e^{j\pi m \Delta\phi_i} + n. \tag{4.50}$$

We first estimate the phase of the received signal, followed by using the estimated phase to estimate the user's angle. According to the previous analysis, for a wideband beam, the first-order approximation of the signal will result in a large error. Therefore, we estimate the AoA based on the signal phase.

$$\min_{\Delta\phi_i} |y - \sum_{m=0}^{M-1} w_m e^{j\pi m \Delta\phi_i}|_F^2. \tag{4.51}$$

The wide beam has a slowly changing phase response, and for a given estimation accuracy, only a few search iterations are required. In addition, similar to the DFT beam, we use Taylor-series expansion to the wide beam phase. Without loss of generality, we just analyze the results of the i-th RF chain. According to the previous derivation, by using the wide beam $\boldsymbol{w}_{wide}$ to construct a beam $\boldsymbol{w}_k$ pointing to θ_k, the signal received can be written as

$$s(\phi_k) = \sum_{m=0}^{M-1} \boldsymbol{w}_k(m) e^{j2\pi f m \frac{d\phi_k}{c}}. \tag{4.52}$$

The first-order Taylor expansion of the signal phase can be expressed as

$$\angle s(\phi_k + \Delta\phi_k) \approx \angle s(\phi_k) + \beta(\phi_k)\Delta\phi_k. \tag{4.53}$$

According to the phase differentiation, we get

$$\beta(\phi_k) = \Im\left(\frac{\partial s(\phi_k)}{\partial \phi_k}\frac{1}{s(\phi_k)}\right) = \Im\left(\frac{W'(\phi_k)}{W(\phi_k)}\right), \tag{4.54}$$

where $W^{'}(\phi_k) = \sum_{m=0}^{M-1}\left(j\pi m \boldsymbol{w}_{wide}(m)\cdot e^{-j2\pi f m\frac{d\phi_k}{c}}\right)$, and $W(\phi_k) = \sum_{m=0}^{M-1}\left(\boldsymbol{w}_{wide}(m)\cdot e^{-j2\pi f m\frac{d\phi_k}{c}}\right)$. We thus get the following result:

When $t = 0$, we get

$$\hat{\phi_k}^{(0)} = \phi_k. \tag{4.55}$$

When $t > 0$, we get

$$\hat{\phi_k}^{(t)} = \hat{\phi_k}^{(t-1)} + \Delta\phi_k, \tag{4.56}$$

where $\Delta\phi_k = \frac{\angle[y_k - \sum_{k'\neq k} s(\hat{\phi_{k'}}^{(t-1)})] - \angle s(\hat{\phi_k}^{(t-1)})}{\beta(\hat{\phi_k}^{(t-1)})}$.

4.4.4 Angle Tracking by Kalman Method

According to the previous analysis, in the presence of the high gain of the large-scale array, a high SNR can be assumed at that time. Under the condition of high SNR, the estimate of θ also gradually obeys the Gaussian distribution. The observation model can thus be rewritten as

$$\begin{aligned}\boldsymbol{\Phi} &= \sin(\boldsymbol{\theta}) + \boldsymbol{n},\\ \boldsymbol{\Psi} &= \arcsin(\boldsymbol{\Phi}) = \boldsymbol{\theta} + \boldsymbol{\eta},\end{aligned} \tag{4.57}$$

where $\boldsymbol{n} \sim N(\boldsymbol{0}, \boldsymbol{\Sigma})$, $\boldsymbol{\eta} \sim N(\boldsymbol{0}, \boldsymbol{\Sigma}^{'})$, $\boldsymbol{\Sigma}$ represents the CRLB in high SNR, and $\Sigma^{'}$ denotes the CRLB after performing $arcsin$. We can thus rewrite the entire tracking model.

State update model is

$$\begin{bmatrix}\boldsymbol{\theta}(n)\\ \boldsymbol{w}(n)\end{bmatrix} = \begin{bmatrix}\boldsymbol{I} & \boldsymbol{I}\\ \boldsymbol{0} & \boldsymbol{I}\end{bmatrix}\begin{bmatrix}\boldsymbol{\theta}(n-1)\\ \boldsymbol{w}(n-1)\end{bmatrix} + \begin{bmatrix}\boldsymbol{u}(n)\\ \boldsymbol{v}(n)\end{bmatrix}. \tag{4.58}$$

Observation model is

$$\boldsymbol{\Psi}(n) = \begin{bmatrix}\boldsymbol{I} & \boldsymbol{0}\end{bmatrix}\begin{bmatrix}\boldsymbol{\theta}(n)\\ \boldsymbol{w}(n)\end{bmatrix} + \boldsymbol{\eta}. \tag{4.59}$$

Therefore, for performing angle tracking in large-scale antenna systems, we construct the system's state update model and observation model through the above methods. By using the modified observation model, we can construct the same structure as traditional Kalman filtering. The entire tracking algorithm can thus be described as Algorithm 4.2.

Algorithm 4.2 Tracking algorithm

```
Require: analogy matrix $G_A$; received signal $y$;
Ensure: estimated angle $\hat{\theta}(n)$; digital matrix $G_D(n)$
Based on the CS structure, using a random analog matrix to estimate the AoA;
Based on the CS estimated angle, using GD method to estimate the angle more accurately;
for $n = 1, 2, \cdots$ do
    Using GAMP method to estimate the angle in n-th frame $\hat{\phi}_i(n), i = 1, 2, \cdots, K$;
    Initialize $\hat{\phi}_i^{(0)}(n)$ by (4.48), beam $i$ is a narrow beam;
    Initialize $\hat{\phi}_i^{(0)}(n)$ by (4.55), beam $i$ is a wide beam;
    while $t \leq Iter_{\max}$ do
        Based on (4.49) and (4.56), using modified EM method to calculate $\hat{\phi}^{(t)}(n)$;
    end while
    Based on estimated angles, $\boldsymbol{\Psi}(n) = \hat{\boldsymbol{\Phi}}^{(t)}(n)$;
    Based on the system model and observation model, using the Kalman filter to calculate
    parameters such as $\hat{\boldsymbol{\theta}}(n|n)$ , $\hat{\boldsymbol{w}}(n|n)$ and $\boldsymbol{G}_D(n) = \left[\boldsymbol{G}_A \boldsymbol{H}(\hat{\boldsymbol{\theta}}(n|n))\right]^{-1}$.
end for
```

4.5 Numerical Results and Analysis

In this section, the proposed algorithm is verified by numerical simulation. First, for the long pilot model in the initial stage, we verify the estimation performance based on the random phase matrix. Secondly, for the preliminary estimation of the AoA in the beam during the tracking process, we perform simulations to verify the estimated performance of the GAMP method. Finally, the performances of the modified EM algorithm and the EKF algorithm are compared in two aspects: instantaneous observation performance and tracking performance. Simulation results show that the observation performance and tracking performance of the proposed algorithm outperform the EKF algorithm in both motion models.

4.5.1 Sparse Channel Estimation in the Initial Stage

In the initial stage, we use an array of $M = 64$ antennas to estimate the AoA of $K = 6$ users simultaneously. Because of the long pilots with orthogonality, at the receiving end, it is possible to both eliminate interference from other users through matching filtering operations and improve the SNR. The joint estimation can be simplified to K independent angle estimations. Figure 4.5 compares the performance of the random matrix and DFT matrix in estimation. It is shown that the performance of the random matrix is much better in most angles, except in those angles to which the DFT beam points. Thanks to the Gaussian noise, the MLE estimation here can reach CRLB gradually. For higher SNR, the noise power becomes smaller and the peak value becomes higher.

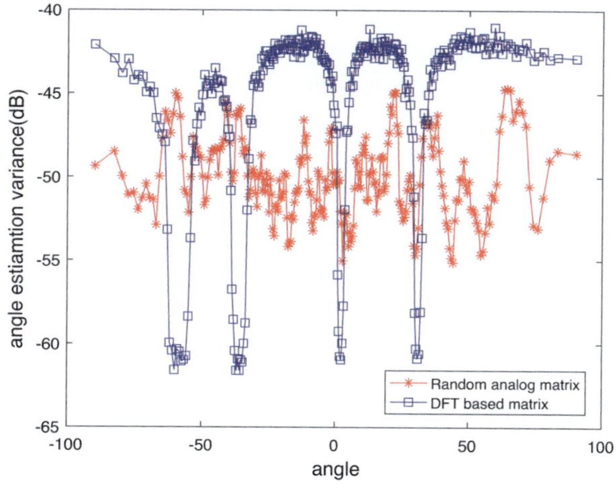

Fig. 4.5 Initial phase, using random phase matrix to estimate the AoA. The random matrix has a smaller variance in angle estimation

4.5.2 GAMP-Based Observations in the Tracking Phase

During the tracking phase, we assign one RF link to each user. For slower users, a beam based on the DFT matrix is allocated, and for faster users, a wide beam is allocated. From Fig. 4.6a, we can find that the beamwidth of the designed wide beam is much larger than that of the DFT. Meanwhile, to achieve a wider beam, we also have to sacrifice the beam gain, corresponding to about 10 dB. In addition, the wide beam has higher side lobes and is more susceptible to interference from other users. In the tracking phase, because the beam is relatively wide, the angle can not be estimated just by indicating the center angle of the beam like that in the DFT

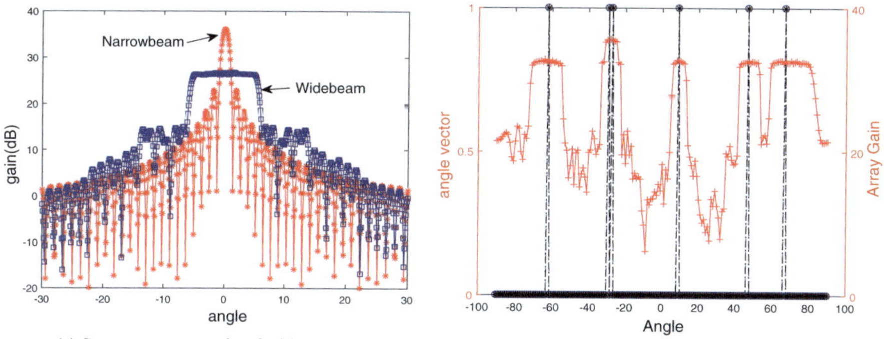

(a) Space spectrum gain of wide beam and narrow beam

(b) Preliminary observations of angles through GAMP

Fig. 4.6 **a** The wide beam has a wider beam width at the cost of a lower power gain. **b** In the tracking process, to deal with the problem that more than one angle has the same gain and phase in wide beam, preliminary observations of angles are performed through GAMP narrow beam. Therefore, we need to use the GAMP method for preliminary AoA estimation. As shown in Fig. 4.6a, the system with an array of 128 antennas serves 6 high-speed users. Two of these users are closer, making their corresponding beams overlap. It can be found through simulation that the GAMP algorithm can estimate the range of the AoA of each user according to the Rayleigh limit of the array. The estimation accuracy is sufficient to ensure the subsequent observation performance of the system.

4.5.3 *Angle Observation Based on Modified EM Algorithm*

The simulation in this section verifies the performance of the modified observation model based on the proposed EM algorithm. Compared with the method of local linearization, we use the relationship between the angle and the phase of the received signal. In the simulation, the number of antennas is $M = 64$, 128, and 256, and the number of users is $K = 6$. Each user corresponds to a RF link, and there is no overlap between beams. We performed simulation verification for the narrow beam and wide beam respectively.

4.5.3.1 Narrow Beam

Figure 4.7a, b show the comparison of the performance between the modified EM algorithm and the traditional extended Kalman model in the case of using a narrow beam to serve slow users. Limited by the accuracy of local linearization, as the SNR increases, the higher-order residuals of Taylor expansion become the main source of

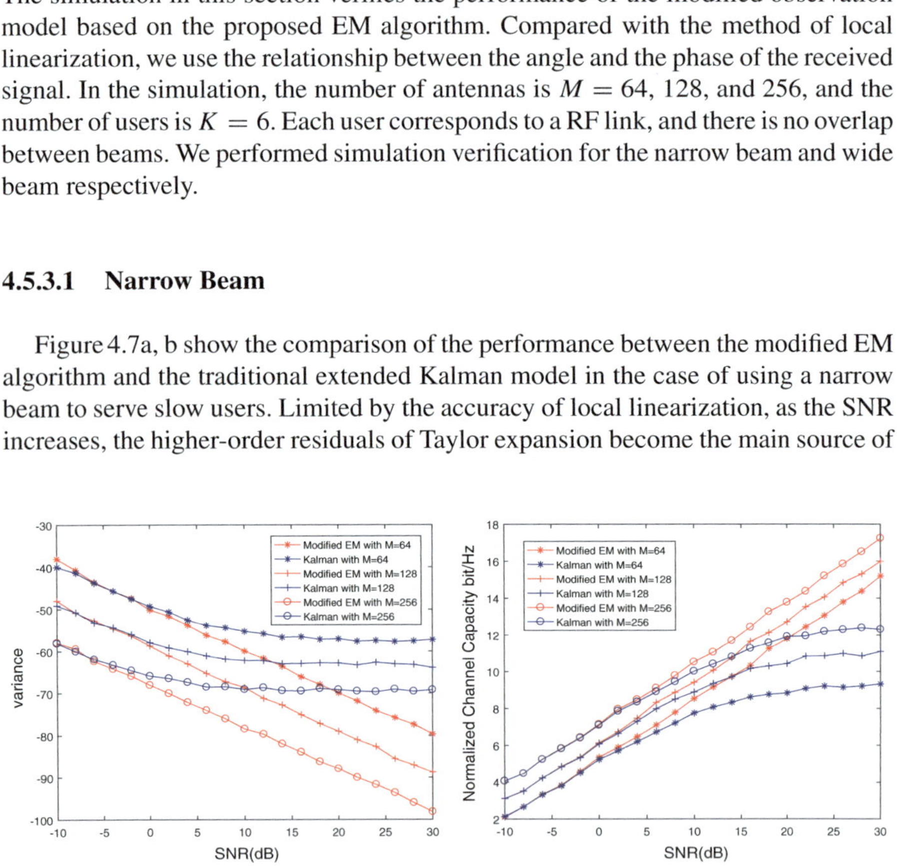

(a) Estimation variance of modified EM method and EKF method in narrow beam

(b) Channel capacity of modified EM method and EKF method in narrow beam

Fig. 4.7 Angle observation based on the modified EM method and EKF method with the DFT beam. (a) The estimation variance of only one observation (b) The channel capacity based on angle observation

estimation errors, and the observation error in extended Kalman model tends to be fixed and no longer decreases. Especially when the angle is at the edge of the beam, the linearization error rises sharply. The modified observation model based on the EM algorithm can avoid the situation successfully. Therefore, Fig. 4.7a, b show that the modified observation model based on the EM algorithm can make full use of the performance gain of high SNR, improving the observation accuracy, and increasing the channel capacity.

4.5.3.2 Wide Beam

Figure 4.8a, b show the comparison of the observation performance between the modified observation model pair based on the EM algorithm and the traditional local linearization model in the case of using a wide beam to serve slow users. Similar to the previous results, the local linearization method under a wide beam is also limited by the Taylor series residual error. Compared with the narrow beam of the DFT matrix, under the wide beam, the estimation error is enhanced by about 10dB as a whole, which is consistent with the loss of the wide beam gain by 10dB relative to the narrow beam. Similarly, because the wide beam gain is low, the corresponding channel capacity is also reduced accordingly.

It is worth noting that the local linearization method has slightly better angular estimation error performance and channel capacity performance in the case of wide beams than in the case of narrow beams. This is because, for a narrow beam, when the AoA is located at the edge of the beam, it will not only cause a change in phase but also a sharp attenuation of amplitude. Compared to the wide beam, it can be seen

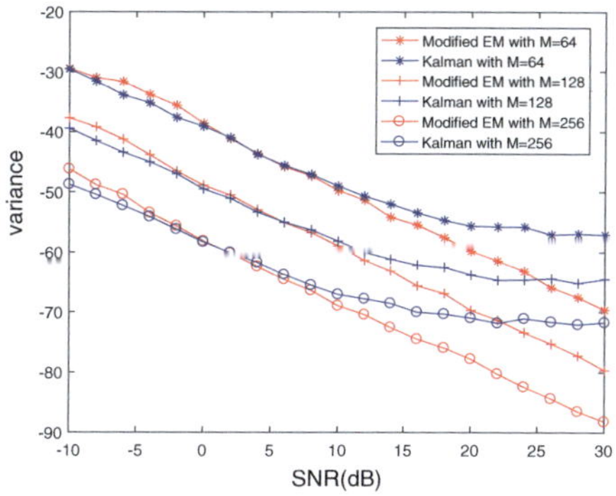

(a) Estimation variance of modified EM method and EKF method in wide beam

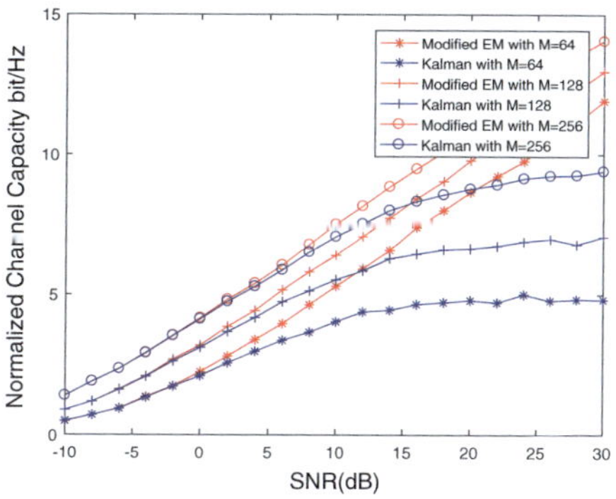

(b) Channel capacity of modified EM method and EKF method in wide beam

Fig. 4.8 Angle observation based on modified EM method and EKF method with the wide beam. (a) The estimation variance of only one observation (b) The channel capacity based on angle observation

from Fig. 4.6a that inside the beam, the gain of the beam is fixed. Therefore, the gain of the wide beam is flatter and more suitable for local linearization operation.

The reason for the floor effect is that, for the AoA estimation, the linear approximation is a biased estimation, the estimation deviation corresponds to the higher-order term of Taylor expansion, and the phase estimation is unbiased. For the case where the noise power is large, since the noise is larger than the estimation deviation, the performance of this method is relatively close. When the noise power is small, the estimated deviation exceeds the noise, and the performance of the EKF is limited by the deviation and enters the step.

4.5.4 AoA Tracking Based on EM Algorithm

The tracking model proposed in this chapter is compared with the traditional EKF model. The number of antennas is 128 and the number of users is 6. Two kinds of motion models are investigated, one is that the angle is moving along a sine trajectory, and the other is that the angle is moving along a straight trajectory. The simulation focuses on the performance of the system with $SNR = 0$ dB, and the simulation is performed for narrow beam slow users and wide beam fast users respectively.

4.5.4.1 Narrow Beam

Aiming at the narrow beam situation, the simulation examined the case where the user has a sine motion model in the range of $-37.7°$ to $-38.7°$ and the entire moving process lasted 500-time slots. Through the simulation result Fig. 4.9a, b, we can find

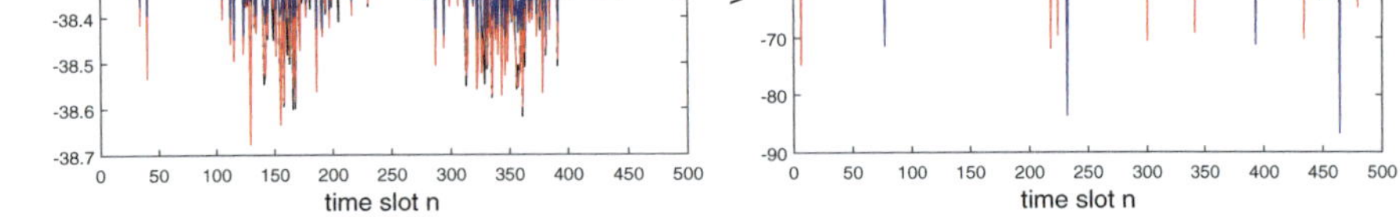

(a) Tracked angle with sine motion in narrow beam

(b) Tracking variance with sine motion in narrow beam

Fig. 4.9 Tracking a sine-moving angle by modified EM method and EKF method with DFT beam. **a** Angle tracked in time series **b** Variance tracked in time series

that both the tracking model provided in this chapter and the traditional EKF model can track the angle. Through Fig. 4.9b, the algorithm provided in this chapter has higher tracking accuracy. By observing Fig. 4.9a, as the user's angle is at the edge of the beam, relative to the proposed algorithm, the tracking error of EKF increases significantly, and when the user's angle is at the center of the beam. The performance of the two is basically the same. In Fig. 4.9a, at the positions of the sine peaks and troughs, the overall estimation result of EKF is relatively small.

Similar to the sine motion model, the simulation in Fig. 4.10a, b examines the situation where the user moves along a straight line in the range of $-37.7°$ to $-38.7°$, and the entire moving process lasts 500-time slots. Consistent with the results in Fig. 4.9, the performance of the proposed algorithm is better than that of EKF at the edge of the beam. At the center of the beam, their performance is basically the same. From Fig. 4.10a, b, compared with sine motion, the estimation accuracy of the algorithm is better. This is because the linear motion is more consistent with the state update model.

4.5.4.2 Wide Beam

Figure 4.11a, b show the situation where a user moves along a sine in a wide range of $-34°$ to $-42°$ under a wide beam. The entire movement process lasts 500-time slots. Similar to the narrow beam, in the case of a wide beam, the proposed algorithm still has higher tracking accuracy, and EKF also faces the problem of linear approximation error. However, compared with narrow beams, the performance of EKF is more stable under wide beams, and the performance of the two algorithms is much closer. Moreover, in Fig. 4.11b, the tracking error of the two algorithms is worse than that of the narrow beam due to the decrease of the beam gain.

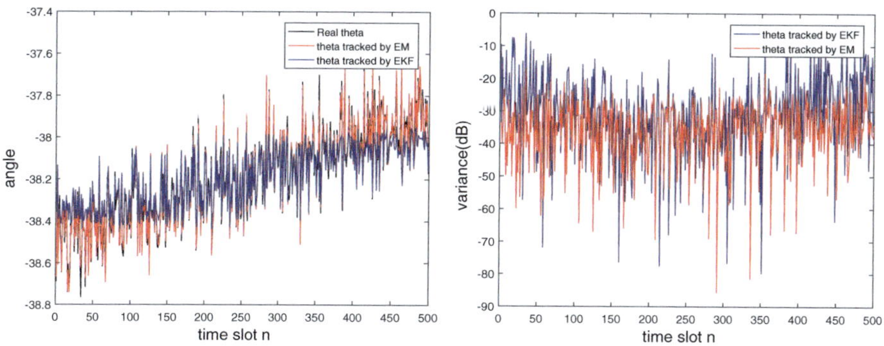

(a) Tracked angle with line motion in narrow beam

(b) Tracking variance with line motion in narrow beam

Fig. 4.10 Tracking a line-moving angle by modified EM method and EKF method with DFT beam. **a** Angle tracked in time series **b** Variance tracked in time series

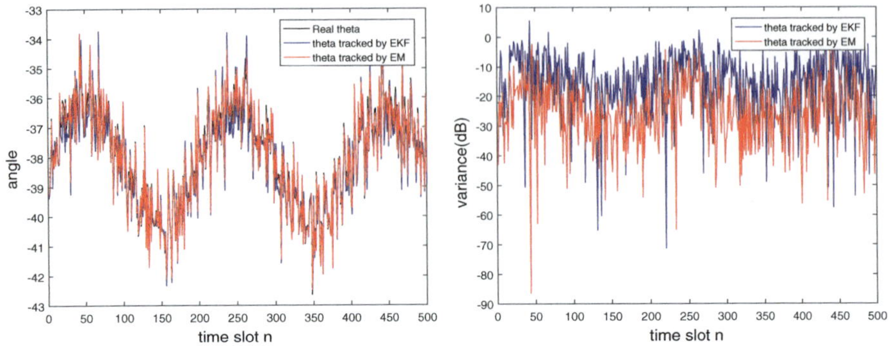

(a) Tracked angle with sine motion in wide beam

(b) Tracking variance with sine motion in wide beam

Fig. 4.11 Tracking a sine-moving angle by modified EM method and EKF method with wide beam. **a** Angle tracked in time series **b** Variance tracked in time series

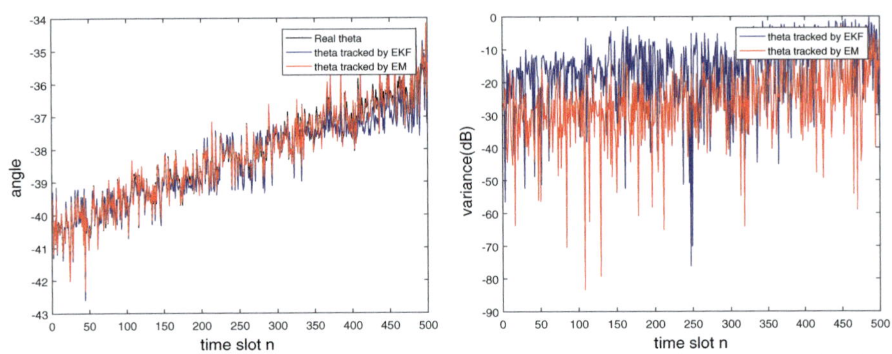

(a) Tracked angle with line motion in wide beam

(b) Tracking variance with line motion in wide beam

Fig. 4.12 Tracking a line-moving angle by modified EM method and EKF method with wide beam. **a** Angle tracked in time series **b** Variance tracked in time series

The simulation in Fig. 4.12a, b examines the situation where the user moves along a straight line in the range of $-34°$ to $-42°$ under a wide beam. As can be seen from Fig. 4.12a, both models can effectively track the AoA change. Compared with the narrow beam, in the wide beam, the performance of this algorithm is similar to that of the EKF, especially when the angle moves to the edge of the beam, the linear approximation in the EKF algorithm can still be close to the true angle. It can be found from Fig. 4.12b that the performance of the algorithm in this chapter is still better than the EKF algorithm. Similar to the narrow beam case, under wide beam, compared with the sinusoidal motion case, the algorithm has better estimation accuracy in the linear motion model.

4.6 Conclusion

In this chapter, we have investigated the angle tracking problem in a THz based large-scale antenna array using a hybrid structure. Aiming at the problem of dimensional disasters with a small number of RF chains, a few observations, and a large angular search space, we have built an initial AoA estimation framework based on a random phase matrix. In the tracking phase, in view of the linearization-error problem in traditional EKF, a preliminary observation of GAMP method has been constructed, followed by a tracking framework proposed based on the modified EM algorithm. Furthermore, relying on phase estimation, an accurate estimation can be achieved with a lower computational complexity. Simulation results showed that the proposed outperform the traditional EKF algorithm in terms of both the estimation performance and tracking accuracy.

References

1. Zhang P, Chen S, Hanzo L (2015) Two-tier channel estimation aided near-capacity MIMO transceivers relying on norm-based joint transmit and receive antenna selection. IEEE Trans Wirel Commun 14(1):122–137
2. Wang Z, Liu W, Qian C, Chen S et al (2017) Two-dimensional precoding for 3-D massive MIMO. IEEE Trans Veh Technol 66(6):5485–5490
3. Zhang J, Chen S, Maunder RG et al (2018) Adaptive coding and modulation for large-scale antenna array-based aeronautical communications in the presence of co-channel interference. IEEE Trans Wirel Commun 17(2):1343–1357
4. Liu W, Wang Z, Cao J et al (2018) Partially-activated conjugate beamforming for LoS massive MIMO communications. IEEE Access 6:56504–56513
5. Pastore A, Joham M, Fonollosa JR (2016) A framework for joint design of pilot sequence and linear precoder. IEEE Trans Inf Theory 62(9):5059–5079
6. Hassibi B, Hochwald BM (2003) How much training is needed in multiple-antenna wireless links. IEEE Trans Inf Theory 49(4):951–963
7. Soysal A, Ulukus S (2010) Joint channel estimation and resource allocation for MIMO systems-part I: single-user analysis. IEEE Trans Wirel Commun 9(2):624–631
8. Soysal A, Ulukus S (2010) Joint channel estimation and resource allocation for MIMO systems-part II: multi-user and numerical analysis. IEEE Trans Wirel Commun 9(2):632–640
9. Gao Z, Dai L, Wang Z et al (2015) Spatially common sparsity based adaptive channel estimation and feedback for FDD massive MIMO. IEEE Trans Signal Process 63(23):6169–6183
10. Xie H, Gao F, Jin S et al (2018) Channel estimation for TDD/FDD massive MIMO systems with channel covariance computing. IEEE Trans Wirel Commun 17(6):4206–4218
11. Xing C, Zhao X, Xu W et al (2019) A framework on hybrid MIMO transceiver design based on matrix-monotonic optimization. IEEE Trans Signal Process 67(13):3531–3546
12. Pan L, Liang L, Xu W et al (2018) Framework of channel estimation for hybrid Analog-and-Digital processing enabled massive MIMO communications. IEEE Trans Commun 66(9):3902–3915
13. Gong Z, Jiang F, Li C (2019) Angle domain channel tracking with large antenna array for high mobility V2I millimeter wave communications. IEEE J Sel Top Signal Process 13(5):1077–1089
14. Zhang Y, Wang D, Wang J et al (2016) Channel estimation for massive MIMO-OFDM systems by tracking the joint angle-delay subspace. IEEE Access 4:10166–10179

15. Brown DR, Wang R, Dasgupta S (2015) Channel state tracking for large-scale distributed MIMO communication systems. IEEE Trans Signal Process 63(10):2559–2571
16. Wang Y, Wang Y, Zhang S et al (2019) Channel tracking and transmission design in 5G large-scale MIMO system. IEEE Access 7:62032–62041
17. Lian L, Liu A, Lau VKN (2019) Exploiting dynamic sparsity for downlink FDD-massive MIMO channel tracking. IEEE Trans Signal Process 67(8):2007–2021
18. Qin Q, Gui L, Gong B et al (2018) Sparse channel estimation for massive MIMO-OFDM systems over time-varying channels. IEEE Access 6:33740–33751
19. Ma J, Zhang S, Li H et al (2019) Sparse Bayesian learning for the time-varying massive MIMO channels: acquisition and tracking. IEEE Trans Commun 67(3):1925–1938
20. Peng W, Li W, Wang W et al (2019) Downlink channel prediction for time-varying FDD massive MIMO systems. IEEE J Sel Top Signal Process 13(5):1090–1102
21. Gao X, Dai L, Zhang Y et al (2017) Fast channel tracking for terahertz beamspace massive MIMO systems. IEEE Trans Veh Technol 66(7):5689–5696
22. Stratidakis G, Ntouni GD, Boulogeorgos AAA et al (2020) A low-overhead hierarchical beam-tracking algorithm for THz wireless systems. Eur Conf Netw Commun 74–78
23. Cheng F, Gui G, Zhao N et al (2019) UAV-relaying-assisted secure transmission with caching. IEEE Trans Commun 67(5):3140–3153
24. Zhao N, Lu W, Sheng M et al (2019) UAV-assisted emergency networks in disasters. IEEE Wirel Commun 26(1):45–51
25. Zhao N, Pang X, Li Z et al (2019) Joint trajectory and precoding optimization for UAV-assisted NOMA networks. IEEE Trans Commun 67(5):3723–3735
26. Rong B, Zhang Z, Zhao X et al (2019) Robust superimposed training designs for MIMO relaying systems under general power constraints. IEEE Access 7:80404–80420
27. Ge Y, Zhang W, Gao F et al (2019) Angle-domain approach for parameter estimation in high-mobility OFDM with fully/partly calibrated massive ULA. IEEE Trans Wirel Commun 18(1):591–607
28. Zhao J, Gao F, Jia W et al (2017) Angle domain hybrid precoding and channel tracking for millimeter wave massive MIMO systems. IEEE Trans Wirel Commun 16(10):6868–6880
29. Haghighatshoar S, Caire G (2018) Low-complexity massive MIMO subspace estimation and tracking from low-dimensional projections. IEEE Trans Signal Process 66(7):1832–1844
30. Qiu T, Fu X, Sidiropoulos ND et al (2018) MISO channel estimation and tracking from received signal strength feedback. IEEE Trans Signal Process 66(7):1691–1704
31. Mai R, Le-Ngoc T (2019) Nonlinear hybrid precoding for coordinated multi-cell massive MIMO systems. IEEE Trans Veh Technol 68(3):2459–2471
32. Al-Shoukairi M, Schniter P, Rao BD (2018) A GAMP-based low complexity sparse Bayesian learning algorithm. IEEE Trans Signal Process 66(2):294–308
33. Zhang Z, Rao BD (2011) Sparse signal recovery with temporally correlated source vectors using sparse Bayesian learning. IEEE J Sel Top Signal Process 5(5):912–926
34. Zhao J, Xie H, Gao F et al (2018) Time varying channel tracking with spatial and temporal BEM for massive MIMO systems. IEEE Trans Wirel Commun 17(8):5653–5666

Chapter 5
Joint Angle Delay Estimation in MIMO Communication Systems

Qi Lu, Xuhui Ding, and Dekang Liu

In this chapter, the joint angle and delay estimation problem of the THz large-scale antenna array is treated. Section 5.1 introduces the research status of joint angle delay estimation in MIMO communication systems. Section 5.2 describes the THz large-scale antenna array structure model and gives the received signal model. In Sect. 5.3, we derive the CRLB on the delay-angle estimation of the system model. Furthermore, we prove the orthogonality of different incident angles and delay signals, and give an estimation method of progressively reaching the CRLB. Section 5.4 proposes a compression structure to reduce the dimension of the signal from two aspects: airspace domain and frequency domain. The solution method based on GAMP is given. Section 5.5 evaluates the performance of the algorithm given by numerical simulation. Finally, conclusions are drawn.

5.1 Introduction

Data synchronization is the core functionality of communication systems, especially for THz high-speed communications [1]. Due to the limitations of technology, THz communication usually uses single-carrier modulation to improve signal power. In the THz communication systems, the signal bandwidth could be so high, that it is

Q. Lu · X. Ding
Beijing Institute of Technology, Beijing, China
e-mail: luqi@bit.edu.cn

X. Ding
e-mail: dingxuhui@bit.edu.cn

D. Liu (✉)
Beihang University, Beijing, China
e-mail: liudekang@buaa.edu.cn

X. Ding et al. (eds.), *Performance Analysis and Improvement in MIMO Communication Systems*, https://doi.org/10.1007/978-981-96-5951-7_5

difficult for ADC and DAC to reach the Nyquist sampling rate and high quantization bit number at the same time [2, 3]. Taking a 300 GHz system as an example, even for a relatively narrowband signal, its bandwidth will still exceed 30 GHz, which is beyond the sampling rate range of current mainstream ADC and DAC. Moreover, at very high sampling rates, the accuracy of ADC and DAC will deteriorate drastically, which is not suitable for high-order modulation. In such circumstances, only simple phase shift keying (PSK) modulation, BPSK modulation, OOK modulation, etc. can be realized [4]. In addition, since the duration of each symbol is very short, extremely high delay estimation accuracy is required. Compared with the OFDM method, the accuracy of delay estimation in single carrier modulation must be well within the interval of one sampling clock [5–7]. Tiny errors can greatly degrade the performance of the system [8, 9].

Considering that the output power of the THz amplifier is still small, and the THz signal is attenuated rapidly, e.g., for a signal at 300 GHz, the distance of 1 meter will cause an attenuation of 82 dB. To guarantee an acceptable receiving SNR, the corresponding array gain is usually higher than 50 dB, and the corresponding beamwidth is about 0.5°. This requires AoA estimation to have an extremely high accuracy [10–12]. Even a small azimuth angle error will seriously damage users' communication quality. At the same time, angle estimation must be performed in a very short time. Considering fast-moving cars or Unmanned-Aerial-Vehicles that have a LoS channel with users, a system with a beamwidth of about 0.5° requires least 60 or even 100 angle estimations within 1 s to ensure that the receiver stays within the beam [13–15]. This undoubtedly requires a high-precision, high-speed estimation method [12, 16–19].

In delay estimation and angle estimation, thanks to the independent identical Gaussian distribution of noise, the MLE is usually adopted [16, 17]. For example, in delay estimation, by utilizing the pilot signal with good autocorrelation property, the optimal delay estimation is fixed to the time at which the peak of the matched filter's response is caught. In angle estimation, the best estimation is usually the angle where the maximum value of the spatial spectrum occurs. However, both estimations' performance is necessarily limited by the grid. To improve the estimation accuracy, more detailed meshing is needed. Since the large-scale THz array itself requires extremely high accuracy of angle estimation and delay estimation, compared with the microwave communication system, the search space will increase by tens or even hundreds of times, which greatly increases the computational complexity. Therefore, on the one hand, the estimator should have low computational complexity. On the other hand, the estimator can achieve CRLB progressively off-grid under high SNR conditions.

For the joint estimation problem of time delay estimation and angle estimation, two ways are helping to reduce the computational complexity.

One approach is to simplify the complexity of angle estimation by using sparsity of angle, such as a sparse array, analog matrix [20–25]. The angle estimation essentially uses the spatial positional relationship of several antennas to estimate the incident angle of the signal, while in a THz based large-scale system, the number of antennas usually reaches several hundred. To achieve a larger array size, if the conventional half-wavelength array is required, the number of antennas and RF channels will exceed an acceptable range, which greatly increases the complexity. For example, in a 300 GHz THz system, the wavelength is 1 mm, and the array spacing must be less than 0.5 mm, which requires an extremely high processing precision. Moreover, the corresponding antenna size is also less than 0.5 mm, which leads to a lower antenna gain. Only in the size of 1 cm $\times$ 1 cm, $20 \times 20 = 400$ THz antennas are deployed. For digital-analog hybrid arrays, at least 400 phase shifters are required, which can be very difficult for system integration. Moreover, even if there is an advantage in the number of antennas, the antenna array gain of 1 cm $\times$ 1 cm size obviously does not satisfy the communication requirement of the THz band. Therefore, To increase the gain of a single antenna, it is necessary to increase the size of a single antenna, which in turn increases the antenna spacing. We can employ a sparse array structure, using high-gain antennas, to arrange arrays in equally spaced sparse arrays.

Another method is to simplify the delay estimator by using sparsity of time delay, such as WMC structure [26–29]. The common delay estimator is to use the FFT technique to achieve the matching filtering process. Therefore, the digital signal processor needs to perform frequency domain transformation on the sampling signal of the ADC. Since the general modulation signal bandwidth can reach tens of GHz, it is rational to consider reducing the signal bandwidth by using the WMC structure. With this structure, the bandwidth can be reduced to several hundred megahertz, so that a normal ADC can support the Nyquist sampling rate. Meanwhile, the complexity of signal processing can be reduced synchronously. The dimensions of the signal can be greatly reduced in two approaches. Furthermore, we could use GAMP-based CS recovery technology [30–35].

Notation: The following notational conventions and definitions are adopted. The normal-faced letters denote scalars, while bold-faced lower-case and upper-case letters denote vectors and matrices, respectively. $\boldsymbol{Z}^{\mathrm{H}}$, $\boldsymbol{Z}^{\mathrm{T}}$, $\mathrm{Tr}(\boldsymbol{Z})$ and $|\boldsymbol{Z}|$ denote the Hermitian transpose, transpose, trace and determinant of $\boldsymbol{Z}$, respectively. $\mathbb{E}\{\cdot\}$ represents statistical expectation. The symbol $\mathrm{diag}\big\{\{\boldsymbol{A}_k\}_{k=1}^{K}\big\}$ denotes the block diagonal matrix whose diagonal sub-matrices are $\boldsymbol{A}_1, \ldots, \boldsymbol{A}_K$. The identity matrix of appropriate dimension is denoted by $\boldsymbol{I}$, and $\otimes$ denotes Kronecker product. In addition, $\mathrm{vec}(\boldsymbol{Y})$ denotes the vectorization operation which stacks the columns of $\boldsymbol{Y}$, and $\|\cdot\|_F$ is the matrix Frobenius norm. $[\boldsymbol{X}]_{:,N}$ consists of the first N columns of $\boldsymbol{X}$, $[\boldsymbol{X}]_{[1:N_1;1:N_2]}$ is the sub-matrix consisting of the first N_1 rows and the first N_2 columns of $\boldsymbol{X}$, and $[\boldsymbol{X}]_{i,j}$ is the (i, j)th element of $\boldsymbol{X}$.

5.2 System Structure and Signal Model

Limited by the immature technology of THz components, the signal power is relatively small. In order to utilize the efficiency of the power amplifier at the transmitting end as much as possible, the constant envelope training pilot is used [18]. The constant modulus phase-coded training pilot containing I symbols is given by

$$x(t) = \sum_{i=0}^{I-1} x_i \cdot p(t - iT_f). \tag{5.1}$$

In which $x_i = \rho e^{j\theta_i}$ represents the ith training symbol. To guarantee constant modulus, we make $\rho = 1$. Furthermore, $p(t)$, representing a shaped pulse for one symbol, could be chosen as a rectangular pulse, a root-raised-cosine pulse, or a Gaussian pulse. The duration of one symbol is T_f. The frequency domain response of this pilot is given by

$$X(f) = FT\{x(t)\} = \sum_{i=0}^{I-1} x_i \cdot P(f) \cdot e^{j2\pi f(i \cdot T_f)}. \tag{5.2}$$

The THz channel response is $h(t - \tau)$, and the frequency response is $H(f)e^{j2\pi f\tau}$, correspondingly [1]. In particular,

$$\begin{aligned} h(t) &= F^{-1}\{H(f)\}, \\ H(f) &= \frac{c}{4\pi f d_T} \cdot exp\left(-\frac{1}{2}K(f)d_T\right), \end{aligned} \tag{5.3}$$

in which $K(f)$ represents the attenuation coefficient in THz frequency f, d_T represents the distance between the BS and the user, and τ represents the time delay from BS to the user, which we need to estimate. The signal received through the THz channel and its spectrum can respectively be deduced as

$$\begin{aligned} y(t) &= x(t) * h(t - \tau) = \sum_{i=0}^{I-1} x_i \cdot g(t - iT_f - \tau), \\ Y(f) &= X(f) \cdot H(f)e^{j2\pi f\tau} = \sum_{i=0}^{I-1} G(f) \cdot e^{j2\pi f(iT_f)} \cdot e^{j2\pi f\tau}. \end{aligned} \tag{5.4}$$

Here, $g(t) = p(t) * h(t)$ represents the pulse after the shaped pulse passes through the THz channel. Its Fourier transform is $G(f)$. Considering that the reflection of the THz signal is very weak in actual conditions, the power of such a multipath signal is much smaller than that of LoS. Therefore, only the signal component of LoS is considered here. At the BS side, we consider a line array receiver equipped with M antennas. If the interval between two adjacent antennas is d, the dimension of this antenna array will be Md. For example, a THz communication system occupies center frequency $f_c = 150\,\text{GHz}$ with wavelength $\lambda = c/f = 2\,\text{mm}$, and the interval between two antennas will be $d = \lambda/2 = 1\,\text{mm}$. Evidently, in a 128 antenna array, the size is $D = Md = 128\,\text{mm}$. Meanwhile, a signal with 10 GHz symbol rate is transmitted, which has a duration of $T_f = 100$ ps. Then,

$$T_f \times c = 100 \text{ ps} \times 3 \times 10^8 \text{ m/s} = 30 \text{ mm} \leq 128 \text{ mm}. \tag{5.5}$$

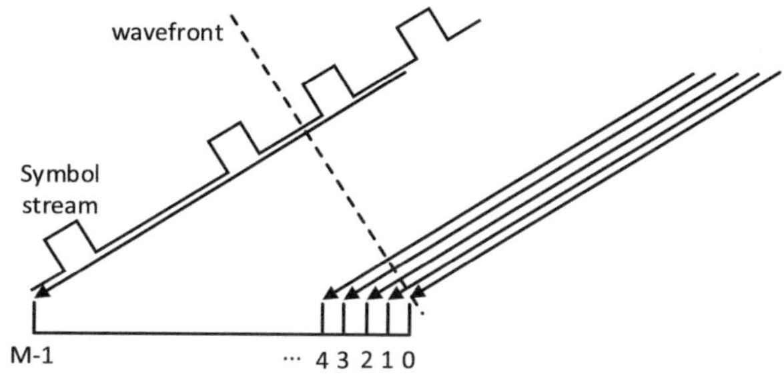

Fig. 5.1 Wideband signal received by a large-scale half-wave array with M antennas

In such occasion, as shown in Fig. 5.1, when the first symbol has just arrived at the No.0 antenna, the fifth symbol is already on the No.127 antenna, where the existing narrowband assumption is no longer satisfied [5, 6]. At this time, the signal in each antenna is no longer a different phase-shifted copy of the same symbol, but a completely different symbol. In this case, we need to rewrite the model of the received signal from the wideband array. Considering the signal arrived at mth antenna is $y_m(t)$ with AoA θ, $y_m(t)$ can be derived as

$$y_m(t) = y\left(t - m\frac{d\sin\theta}{c} - \tau\right) \times e^{j2\pi f_c(m\frac{d\sin\theta}{c}+\tau)} + w_m(t). \tag{5.6}$$

Here, $y(t - m\frac{d\sin\theta}{c} - \tau)$ represents the received baseband signal in mth antenna, which is different from the traditional narrowband array where $y(t - m\frac{d\sin\theta}{c} - \tau) \approx y(t-\tau)$. And $e^{j2\pi f_c(m\frac{d\sin\theta}{c}+\tau)}$ represents an additional carrier phase, which is the same as the traditional narrowband array. Then, in the mth antenna, the received signal is

$$y_m(t) = \sum_{i=0}^{I-1} x_i \cdot g\left(t - iT_f - m\frac{d\sin\theta}{c} - \tau\right) e^{j2\pi f_c(m\frac{d\sin\theta}{c}+\tau)} + w(t). \tag{5.7}$$

Correspondingly, we have

$$Y_m(f) = Y(f)e^{j2\pi(f+f_c)m\frac{d\sin\theta}{c}} e^{j2\pi f\tau} + W_m(f). \tag{5.8}$$

Discretizing the spectrum in separate frequency

$$Y_{m,k}(\theta,\tau) = Y_m(f_k) = H_k X_k e^{j2\pi(f_k+f_c)m\frac{d\sin\theta}{c}} e^{j2\pi f_k\tau} + W_m(f_k). \tag{5.9}$$

Its matrix form can be written as

$$\boldsymbol{Y} = \begin{bmatrix} Y_{0,0} & Y_{0,1} & \cdots & Y_{0,K-1} \\ Y_{1,0} & Y_{1,1} & \cdots & Y_{1,K-1} \\ \vdots & \vdots & \ddots & \vdots \\ Y_{M-1,0} & Y_{M-1,1} & \cdots & Y_{M-1,K-1} \end{bmatrix} = \boldsymbol{A}(\theta, \tau)\boldsymbol{H}\boldsymbol{X} + \boldsymbol{W}. \tag{5.10}$$

In which, $[\boldsymbol{A}(\theta, \tau)]_{m,k} = e^{j2\pi(f_k+f_c)m\frac{d\sin\theta}{c}} e^{j2\pi f_k \tau}$, $\boldsymbol{H}\boldsymbol{X} = diag([H_k X_k])$, $[\boldsymbol{W}]_{m,k} = W_m(f_k)$, As described above, the size of the antenna array is much larger than the length of a symbol propagating in space, and as a result, the conventional array model cannot be satisfied. Therefore, we need to adopt a new method of array signal processing.

According to the expression of the received signal, noting that the noise on each antenna and each frequency is independent and identically cyclic Gaussian distributed, we can easily associate the MLE as

$$\ln\left[p(\boldsymbol{Y}\|\theta, \tau)\right] = C - ||\boldsymbol{Y} - \boldsymbol{A}(\theta, \tau)\boldsymbol{H}\boldsymbol{X}||_F^2, \tag{5.11}$$

where C is a constant. The estimation can be represented as an optimization problem like

$$\max_{\theta,\tau} -||\boldsymbol{Y} - \boldsymbol{A}(\theta, \tau)\boldsymbol{H}\boldsymbol{X}||_F^2. \tag{5.12}$$

In the following, we will give a specific description of the MLE method for the unique expression of this problem. Meanwhile, we demonstrate how to eliminate multiple peaks in angle estimation caused by sparse arrays. Then an unambiguous estimation method progressively reaching CRLB is given.

5.3 CRLB of Joint Angle Delay Estimation

Based on the expression of the received signal, we can get the likelihood function.

$$\begin{aligned} Y_{m,k}(\theta, \tau) = Y_m(f_k) &= H_k X_k e^{j2\pi(f_k+f_c)m\frac{d\sin\theta}{c}} e^{j2\pi f_k \tau} + W_{m,k} \\ &= S_{m,k}(\theta, \tau) + W_{m,k}, \end{aligned} \tag{5.13}$$

in which $S_{m,k}(\theta, \tau) = H_k X_k e^{j2\pi(f_k+f_c)m\frac{d\sin\theta}{c}} e^{j2\pi f_k \tau}$ and $W_{m,k} \sim CN(0, \sigma_n^2)$.

The likelihood function is expressed as

$$p(\boldsymbol{Y}, \theta) = \prod_{m,k} \left[\frac{1}{\sqrt{2\pi\sigma_n^2}} e^{-\frac{1}{2\sigma_n^2}|Y_{m,k} - S_{m,k}|^2} \right]. \tag{5.14}$$

Thus, based on the likelihood function of the received signal, we can calculate the Fisher information matrix as

$$\boldsymbol{I} = \begin{bmatrix} I_{\theta,\theta} \ I_{\theta,\tau} \\ I_{\tau,\theta} \ I_{\tau,\tau} \end{bmatrix}. \tag{5.15}$$

According to the characteristics of Gaussian white noise, we get

$$\begin{aligned} \frac{\partial S_{m,k}(\theta,\tau)}{\partial \tau} &= \frac{\partial \left[H_k X_k e^{j2\pi(f_k+f_c)m\frac{d\sin\theta}{c}} e^{j2\pi f_k \tau} \right]}{\partial \tau} \\ &= j2\pi f_k H_k X_k e^{j2\pi(f_k+f_c)m\frac{d\sin\theta}{c}} e^{j2\pi f_k \tau} \\ &= j2\pi f_k S_{m,k}(\theta,\tau), \end{aligned} \tag{5.16}$$

$$\begin{aligned} \frac{\partial S_{m,k}(\theta,\tau)}{\partial \theta} &= \frac{\partial \left[H_k X_k e^{j2\pi(f_k+f_c)m\frac{d\sin\theta}{c}} e^{j2\pi f_k \tau} \right]}{\partial \theta} \\ &= j2\pi(f_k+f_c) H_k X_k e^{j2\pi(f_k+f_c)m\frac{d\sin\theta}{c}} e^{j2\pi f_k \tau} m\frac{d\cos\theta}{c} \\ &= j2\pi(f_k+f_c) S_{m,k}(\theta,\tau) m\frac{d\cos\theta}{c}, \end{aligned} \tag{5.17}$$

$$\begin{aligned} I_{\theta,\theta} &= \frac{1}{\sigma_n^2}\sum_{m,k} |\frac{\partial S_{m,k}(\theta,\tau)}{\partial \theta}|^2 = \frac{1}{\sigma_n^2}\sum_{m,k} |H_k X_k|^2 \left(2\pi(f_k+f_c)m\frac{d\cos\theta}{c}\right)^2, \\ I_{\tau,\tau} &= \frac{1}{\sigma_n^2}\sum_{m,k} |\frac{\partial S_{m,k}(\theta,\tau)}{\partial \theta}|^2 = \frac{1}{\sigma_n^2}\sum_{m,k} |H_k X_k|^2 (2\pi f_k)^2, \\ I_{\theta,\tau} &= \frac{1}{\sigma_n^2}\sum_{m,k} \left(\frac{\partial S_{m,k}(\theta,\tau)}{\partial \theta}\right)^H \left(\frac{\partial S_{m,k}(\theta,\tau)}{\partial \tau}\right) \\ &= \frac{1}{\sigma_n^2}\sum_{m,k} |H_k X_k|^2 (2\pi f_k) 2\pi(f_k+f_c)\left(m\frac{d\cos\theta}{c}\right), \\ I_{\tau,\theta} &= \frac{1}{\sigma_n^2}\sum_{m,k} \left(\frac{\partial S_{m,k}(\theta,\tau)}{\partial \tau}\right)^H \left(\frac{\partial S_{m,k}(\theta,\tau)}{\partial \theta}\right) \\ &= \frac{1}{\sigma_n^2}\sum_{m,k} |H_k X_k|^2 (2\pi f_k) 2\pi(f_k+f_c)\left(m\frac{d\cos\theta}{c}\right), \\ \boldsymbol{I} &= \frac{1}{\sigma_n^2}\sum_{m,k} |H_k X_k|^2 \begin{bmatrix} (2\pi f_k)^2 & (2\pi)^2 f_k(f_k+f_c)m\frac{d\cos\theta}{c} \\ (2\pi)^2 f_k(f_k+f_c)m\frac{d\cos\theta}{c} & (2\pi(f_k+f_c)m\frac{d\cos\theta}{c})^2 \end{bmatrix}. \end{aligned} \tag{5.18}$$

Finally, we get

$$\begin{aligned} var(\theta) &\geq \left[\boldsymbol{I}^{-1}\right]_{\theta,\theta} = \frac{I_{\tau,\tau}}{I_{\theta,\theta}I_{\tau,\tau} - I_{\theta,\tau}I_{\tau,\theta}}, \\ var(\tau) &\geq \left[\boldsymbol{I}^{-1}\right]_{\tau,\tau} = \frac{I_{\theta,\theta}}{I_{\theta,\theta}I_{\tau,\tau} - I_{\theta,\tau}I_{\tau,\theta}}. \end{aligned} \tag{5.19}$$

Here, we can find that the essence of the Fisher information matrix in the CRLB derivation is the weighted sum of different frequency components. In THz communication, we use BPSK modulation, so the signal spectrum is similar to the Sinc function, and the frequency interval can be chosen arbitrarily. Moreover, because the frequency response of the THz channel is frequency-selective, the signal power of different frequencies is different. If an OFDM system is used, the signal power on the subcarriers can be guaranteed to be the same, which will greatly simplify the system model and simplify the derivation of CRLB. The CRLB in OFDM systems can be presented in a more concise form [5, 6].

5.3.1 Joint Angle Delay Estimation Method

Considering that the noise on each component of the received signal is independently and identically distributed, we can rewrite the MLE problem into

$$\min_{\theta,\tau} \sum_{m,k} |Y_{m,k} - S_{m,k}(\theta,\tau)|^2. \tag{5.20}$$

A straightforward approach is to separate θ and τ into a number of non-overlapping grids, and then search each grid point. To simplify the searching problem, we will first prove that the received signals at different grid points are approximately orthogonal. Therefore, the search problem can be expressed as a matched filtering problem of the signal, and only a peak appears at the true θ and τ point. Then, by using the unique structure of the received signal, a simplified method based on modified FFT is presented. According to the regularity of the Gaussian distribution, as long as the mesh is fine enough, the estimation can be guaranteed to reach CRLB progressively. However, finer meshing will increase the computational complexity quickly. Therefore, on the basis of mesh meshing, a meshless progressive CRLB estimation method is proposed, and it is proved that the estimation method can gradually reach CRLB as the SNR increases.

5.3.2 *Proof of Orthogonality*

In this section, we will give a detailed description of how signals with different angles or delays are approximately orthogonal. For simplicity, we first make some definitions: $\phi = \sin\theta$, $f_k = k\Delta f$ and $d = \frac{c}{2f_c}$.

By defining the first signal as $\boldsymbol{S}_1 = \boldsymbol{S}(\phi, \tau)$ and the second signal as $\boldsymbol{S}_2 = \boldsymbol{S}(\phi + \Delta\phi, \tau + \Delta\tau)$, we get

$$\begin{aligned} Tr\left(\boldsymbol{S}_1^H \boldsymbol{S}_2\right) &= Tr\left([\boldsymbol{S}(\phi,\tau)]^H\left[\boldsymbol{S}(\phi+\Delta\phi,\tau+\Delta\tau)\right]\right) \\ &= Tr\left([\boldsymbol{A}(\phi,\tau)\boldsymbol{H}\boldsymbol{X}]^H\left[\boldsymbol{A}(\phi+\Delta\phi,\tau+\Delta\tau)\boldsymbol{H}\boldsymbol{X}\right]\right) \\ &= \sum_m\sum_k |H_k X_k|^2 e^{-j2\pi(f_k+f_c)(m\frac{d\Delta\phi}{c})} e^{j2\pi f_k\Delta\tau}. \end{aligned} \tag{5.21}$$

By define $\Delta\phi = \frac{l}{M}$ and $\Delta\tau = \frac{i}{K\Delta f}$, we have

$$Tr\left(\boldsymbol{S}_1^H \boldsymbol{S}_2\right) = \sum_k |H_k X_k|^2 e^{j2\pi\frac{ik}{K}} \sum_m e^{-j2\pi\left(1+\frac{k\Delta f}{f_c}\right)\frac{ml}{2M}}. \tag{5.22}$$

- When $l = 0, i = 0$:

It is obvious that all signals are combined constructively without waste of energy.

$$Tr\left(\boldsymbol{S}_1^H \boldsymbol{S}_2\right) = M\sum_k |H_k X_k|^2. \tag{5.23}$$

In this case, the received signal power is maximized.

- When $0 < l < M$:

We observe the last summation, in which case the bandwidth of the signal is relatively narrow relative to the carrier frequency. We have $\left(1 + \frac{k\Delta f}{f_c}\right) \approx 1$, and this summation can be simplified as

$$\sum_m e^{-j2\pi\left(1+\frac{k\Delta f}{f_c}\right)\frac{ml}{2M}} \approx \sum_m e^{-j2\pi\frac{ml}{2M}} = 0. \tag{5.24}$$

Since this result is true for each frequency component f_k, the signals with different angles of incidence are approximately orthogonal to each other in general.

- When $l = 0, 0 < i < K$:

Since the latter summation is a constant, we only need to observe the previous summation. The former summation number can be regarded as the autocorrelation of the received pilot signal. Obviously, under the condition that the pseudorandom signal is used as the pilot signal, the autocorrelation function reaches the maximum value only when $i = 0$. When $0 < i < K$, the autocorrelation function of the signal is approximately 0, leading to

$$\sum_k |H_k X_k|^2 e^{-j2\pi \frac{ik}{K}} \approx 0. \tag{5.25}$$

Thus, we demonstrate that the observed signals with different angles and delays are orthogonal. Intuitively, the observed signals with different angles of incidence and time delay are orthogonal. On the one hand, for signals on a certain frequency component, the signals entering the array from different incident angles are orthogonal because the array size is relatively large. On the other hand, since the signals are in the form of random sequences, the signals for different delays are approximately orthogonal. The orthogonality of the signal can also be found through simulation in Fig. 5.2. Only the origin one has a peak, and the result in other cases approaches zero.

5.3.3 FFT-Based ML Estimator

According to the orthogonal form of the signal, we can construct a joint angle delay estimation method based on meshing. In one word, we can mesh the angle of incidence and time delay, and then calculate the joint angle and delay spectrum of the received signal on the grid point. The optimization problem can be written as

$$\max_{\phi,\tau} ||Tr\left(A(\phi, \tau)^H Y\right)||_F^2. \tag{5.26}$$

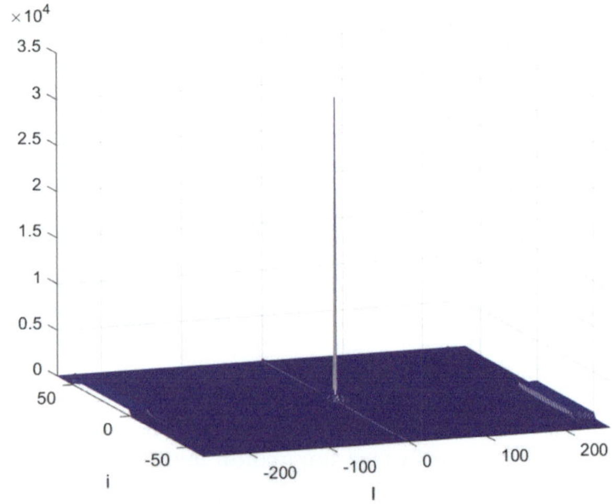

Fig. 5.2 Proof of orthogonality in half-wave array (signals with nonzero i or l are approximately orthogonal)

However, it is obvious that the signal dimensions are very high, due to both the large bandwidth and the large size of the array. Moreover, according to the previous analysis, the large-scale THz communication system requires high angle estimation accuracy and high delay estimation accuracy. As a result, the space of meshes that need to be searched increases accordingly. In this way, a simple search leads to huge computational complexity. This is unacceptable for communication systems that require a fast estimation.

By reviewing the form of the received signal

$$\begin{aligned} Y_{m,k} &= H_k X_k e^{j2\pi(f_k+f_c)m\frac{d\sin\theta}{c}} e^{j2\pi f_k \tau} + w_{m,k} \\ &= H_k X_k \left[e^{j2\pi f_k \tau}\right]\left[e^{j2\pi(f_k+f_c)m\frac{d\sin\theta}{c}}\right] + w_{m,k}, \end{aligned} \tag{5.27}$$

we can find that the received signal can be divided into three parts:

- H_k and X_k represent channel and data, respectively.
- $\left[e^{j2\pi f_k \tau}\right]$ represents an additional phase in each frequency caused by time delay.
- $\left[e^{j2\pi(f_k+f_c)m\frac{d\sin\theta}{c}}\right]$ represents additional phase in each frequency caused by AoA.

In the definitions, only the third item is related to the distribution of the antenna array. For a specific frequency component, we can perform Fourier transformation on $\left[e^{j2\pi(f_k+f_c)m\frac{d\sin\theta}{c}}\right]$ to search angle.

For a specific frequency component f_k, we can perform discrete Fourier transformation on signals received in M antennas.

$$\begin{aligned} R(k) &= \sum_{m=0}^{M-1} e^{-j2\pi\frac{km}{M}} e^{j2\pi(f_k+f_c)m\frac{d\sin\theta}{c}} \\ &= \sum_{m=0}^{M-1} e^{-j2\pi\frac{km}{M}} e^{j2\pi\left(1+\frac{f_k}{f_k}\right)m\frac{\sin\theta}{2}}. \end{aligned} \tag{5.28}$$

It is worth noting that the estimated space spectrum should be adjusted by a frequency factor $(1+\frac{f_k}{f_c})$. Indeed, the true AoA is $\sin\theta = \frac{2k}{M\left(1+\frac{f_k}{f_c}\right)}$.

After performing adjustment, the result can be expressed as

$$r_{\theta,k} = \sum_m Y_{m,k} e^{-j2\pi(f_k+f_c)\frac{d\sin\theta}{c}m}. \tag{5.29}$$

In realization, the angle is usually divided into several separate nonoverlapping segments, in which $\sin\theta \in (-1,+1)$.

In this way, we get a joint spectrum of frequency-space. Each column in the joint spectrum represents a spatial spectrum estimate under a certain frequency component.

At this point, we can infer that for a single user, the signal occurs only in the row corresponding AoA of this user, implying that only one row corresponding to the user signal angle is nonzero. The other rows correspond to noise.

We can continue to search for the time delay through the frequency-space spectrum.

$$r_{\theta,\tau} = \sum_k r_{\theta,k} \times (H_k X_k)^* e^{-2\pi f_k \tau}. \tag{5.30}$$

This is equivalent to performing a matched filtering operation on the signal, which can be realized by using FFT.

By employing a modified FFT and a standard FFT, we implement a space-time two-dimensional search of the signal. Through simulation, we can find that in the spatial frequency combined spectrum, there are only spikes at the positions corresponding to the real signals, while other locations have no spikes.

5.3.4 *Progressively Reaching CRLB Estimation with Single User*

According to the previous analysis, we already know that accurate estimation of angle and delay can be obtained by MLE estimation because the signal has good orthogonality of time and space. However, based on the derivation, we can also find that whether the search is based on a grid or FFT to reduce the search complexity, the accuracy of the estimation depends on the accuracy of the grid. In general, the accuracy of the delay estimate is limited by the sampling interval of the ADC, while the estimation accuracy of the angle is restricted in the Rayleigh limit of the array dimension. Therefore, even in high SNR conditions, it is necessary to refine the division of a grid to obtain a higher accurate estimation, for example, to increase the ADC sampling rate and increase the array size. In response to this problem, we still start from the expression of the signal, and use the influence of angle and delay on the array response, followed by proposing an estimation method that can gradually reach CRLB without relying on mesh division.

5.3.4.1 Progressively Reaching CRLB Estimation in Phase Estimation

By revisiting the signal representation, we can see that the estimated θ and τ are in the phase of the signal. Therefore, intuitively, as long as the phase spectrum in the signal spectrum can be accurately estimated, both time delay and angle of incidence can be obtained.

First, for a fixed m and k, we make definitions $\alpha = [2\pi(f_k + f_c)m\frac{d\sin\theta}{c} + 2\pi f_k\tau]$ and $\rho = |H_k X_k|$. In this way, the original signal representation can be rewritten as

$$Y_{m,k} = \rho e^{j[\alpha + \angle(H_k X_k)]} + w_{m,k}, \tag{5.31}$$

in which $\angle(H_k X_k)$ represents a phase of the channel and signal $H_k X_k$, is a known constant. Therefore, in order to estimate α, we will think of using the $arctan(x)$ function to calculate the phase of the complex signal. The estimation of α can be written as

$$\hat{\alpha} = \arctan\left(\frac{\Im(Y_{m,k})}{\Re(Y_{m,k})}\right) - \angle(H_k X_k) = \arctan\left(\frac{\rho\sin(\alpha) + \epsilon_s}{\rho\cos(\alpha) + \epsilon_c}\right) = g_{m,k}(\epsilon_s, \epsilon_c), \tag{5.32}$$

in which $\epsilon_s = \Im(w_{m,k})e^{-j\angle(H_k X_k)}$ and $\epsilon_c = \Re(w_{m,k})e^{-j\angle(H_k X_k)}$, corresponding to noise on the quadrature component and noise on the in-phase component, respectively. Under the condition that noise power is relatively low, we can use one-order Taylor series approximate $g_{m,k}(\epsilon_s, \epsilon_c)$:

$$\begin{aligned} g_{m,k}(\epsilon_s, \epsilon_c) &= \arctan\left(\frac{\rho\sin(\alpha) + \epsilon_s}{\rho\cos(\alpha) + \epsilon_c}\right) \\ &\approx \alpha + \frac{\partial g_{m,k}(\epsilon_s, \epsilon_c)}{\epsilon_s}\epsilon_s + \frac{\partial g_{m,k}(\epsilon_s, \epsilon_c)}{\epsilon_c}\epsilon_c \\ &= \alpha + \frac{\cos\alpha}{\rho}\epsilon_s + \frac{\sin\alpha}{\rho}\epsilon_c. \end{aligned} \tag{5.33}$$

This means that, under low noise conditions, the estimation of phase $g_{m,k}(\epsilon_s, \epsilon_c)$ has a Gaussian distribution like

$$g_{m,k}(\epsilon_s, \epsilon_c) \sim N\left(\alpha, \frac{\sigma_n^2}{|H_k X_k|^2}\right). \tag{5.34}$$

The expectation of the Gaussian distribution represents the true phase. Considering that ϵ_s and ϵ_c are independent, we can combine the two noises into one noise. In this way, we can write the phase obtained by the arctan function into the following form.

For the sake of simplicity, we define $\phi = \sin\theta$ and $\sigma_k = \frac{\sigma_n}{|H_k X_k|}$.
Thus, the phase estimation model can be written as

$$\begin{aligned} \hat{\alpha}_{m,k} &= 2\pi(f_k + f_c)m\frac{d\sin\theta}{c} + 2\pi f_k\tau + \eta_{m,k} \\ &= 2\pi(f_k + f_c)m\frac{d\phi}{c} + 2\pi f_k\tau + \eta_{m,k}, \end{aligned} \tag{5.35}$$

in which $\eta_{m,k} \sim N\left(0, \sigma_k^2\right)$.

5.3.4.2 Linear Phase Combination Estimation

By observing the phase estimation model, we find that this is a typical linear observation MLE problem, so we can write $\hat{\alpha}_{m,k}$ into a vector form as

$$\hat{\boldsymbol{\alpha}} = \boldsymbol{F}\begin{bmatrix}\tau\\ \phi\end{bmatrix} + \boldsymbol{\eta}, \tag{5.36}$$

in which

$$\hat{\boldsymbol{\alpha}} = vec\left(\begin{bmatrix}\hat{\alpha}_{0,0} & \hat{\alpha}_{0,1} & \cdots & \hat{\alpha}_{0,K-1}\\ \hat{\alpha}_{1,0} & \hat{\alpha}_{1,1} & \cdots & \hat{\alpha}_{1,K-1}\\ \vdots & \vdots & \ddots & \vdots\\ \hat{\alpha}_{M-1,0} & \hat{\alpha}_{M-1,1} & \cdots & \hat{\alpha}_{M-1,K-1}\end{bmatrix}\right), \tag{5.37}$$

$$\boldsymbol{\eta} = vec\left(\begin{bmatrix}\eta_{0,0} & \eta_{0,1} & \cdots & \eta_{0,K-1}\\ \eta_{1,0} & \eta_{1,1} & \cdots & \eta_{1,K-1}\\ \vdots & \vdots & \ddots & \vdots\\ \eta_{M-1,0} & \eta_{M-1,1} & \cdots & \eta_{M-1,K-1}\end{bmatrix}\right). \tag{5.38}$$

It is worth noting that $vec(\boldsymbol{Y})$ denotes the vectorization operation which stacks the columns of $\boldsymbol{Y}$.

$$\boldsymbol{F} = \left[\begin{bmatrix}2\pi f_0\\ 2\pi f_1\\ \vdots\\ 2\pi f_{K-1}\end{bmatrix} \otimes \begin{bmatrix}1\\ 1\\ \vdots\\ 1\end{bmatrix}_{M\times 1} \quad \frac{d}{c}\begin{bmatrix}2\pi(f_0 + f_c)\\ 2\pi(f_1 + f_c)\\ \vdots\\ 2\pi(f_{K-1} + f_c)\end{bmatrix} \otimes \begin{bmatrix}0\\ 1\\ \vdots\\ M-1\end{bmatrix}_{M\times 1}\right]. \tag{5.39}$$

Since the individual noises are independent, the noise covariance matrix can be written as

$$\boldsymbol{\Sigma}_n = diag([\sigma_0^2 \mathbf{1}_{1\times M}, \sigma_1^2 \mathbf{1}_{1\times M}, \ldots, \sigma_{K-1}^2 \mathbf{1}_{1\times M}]). \tag{5.40}$$

Thus, the optimal estimate of the parameters can be written as

$$\begin{bmatrix}\hat{\tau}\\ \hat{\phi}\end{bmatrix} = \left(\boldsymbol{F}^T \boldsymbol{\Sigma}_n^{-1} \boldsymbol{F}\right)^{-1} \boldsymbol{F}^T \boldsymbol{\Sigma}_n^{-1} \hat{\boldsymbol{\alpha}}. \tag{5.41}$$

The corresponding estimated variance matrix is

$$\boldsymbol{I}(\hat{\tau}, \hat{\phi}) = \left(\boldsymbol{F}^T \boldsymbol{\Sigma}_n^{-1} \boldsymbol{F}\right). \tag{5.42}$$

Considering $\phi(\theta) = \sin\theta$, the estimated errors of τ,θ are

$$\boldsymbol{I}(\hat{\tau}, \hat{\theta}) = \begin{bmatrix}1 & 0\\ 0 & \frac{\partial\phi}{\partial\theta}\end{bmatrix} \left(\boldsymbol{F}^T \boldsymbol{\Sigma}_n^{-1} \boldsymbol{F}\right) \begin{bmatrix}1 & 0\\ 0 & \frac{\partial\phi}{\partial\theta}\end{bmatrix}, \tag{5.43}$$

$$
\begin{aligned}
var(\theta) &\geq [\boldsymbol{I}(\hat{\tau},\hat{\theta})]^{-1}_{\hat{\theta},\hat{\theta}} = [\boldsymbol{I}(\hat{\tau},\hat{\phi})]^{-1}_{\hat{\phi},\hat{\phi}}\left(\frac{\partial\theta}{\partial\phi}\right)^2 = [\boldsymbol{I}(\hat{\tau},\hat{\phi})]^{-1}_{\hat{\phi},\hat{\phi}}\left(\frac{1}{\cos\theta}\right)^2, \\
var(\tau) &\geq [\boldsymbol{I}(\hat{\tau},\hat{\theta})]^{-1}_{\hat{\tau},\hat{\tau}} = [\boldsymbol{I}(\hat{\tau},\hat{\phi})]^{-1}_{\hat{\tau},\hat{\tau}}.
\end{aligned}
\tag{5.44}
$$

Here, it should be emphasized that the matrix in formula (5.43) and the Fisher information matrix in formula (5.18) in the CLRB derived earlier are exactly the same. The variance of parameters of angle and delay is obtained by performing the inverse of the matrix in (5.43) which is similar to the derivation of CRLB. Therefore, we have proved that the results of the formula (5.44) and formula (5.19) are exactly the same. Therefore, by comparison, we find that the linear phase combination method can ensure that the estimation result reaches CRLB.

5.4 Estimation Method Based on Compressed Sensing Structure

The structured MLE estimation method is given. The angle and delay estimation progressively reaching CRLB can be realized by the grid search method, modified FFT search method, and phase estimation. Moreover, due to the use of the FFT method and the phase combination method, the search step is simplified on the one hand, and the meshing requirement is relaxed on the other hand, which greatly reduces the computational complexity. However, these two approaches are still trapped in the high dimension of the signal. Since the number of antennas in a THz array is usually up to several hundred and the signal bandwidth can reach tens of GHz, a large amount of hardware overhead and computation is required. Therefore, in this section, for the problem of a large number of THz large-scale array antennas and large signal bandwidth, our response gives an estimation method of CS. The method can ensure that the angle estimate reaches the Rayleigh limit of the original array size under the condition of good SNR, and the delay estimation can reach the Rayleigh limit of the signal bandwidth.

5.4.1 Sparse Uniform Linear Array

Consider a THz wideband array with M antennas. The spacing of the antennas is $d \leq \lambda/2$. Considering the traditional large-scale antenna array, in order to achieve the same array gain, the number of antennas often reaches several hundred, and the gain of a single antenna array element is limited by size (may only be 0 dB). In this system, due to the antenna spacing being increased, in order to achieve the same size, only a few high gain (15 dB) antennas are needed to simplify the complexity of the system design.

Taking the 300 GHz band as an example, the signal wavelength is 1 mm. A line array containing 256 elements with a size of 128 mm and an antenna spacing of 0.5 mm is considered, each antenna element having a size of 0.5 mm and a gain of 0 dB. The total gain of the array is 24 dB. Using the sparse array structure of this chapter, the array size is unchanged (still 128 mm) and the antenna spacing is increased to 16 mm, thus requiring only 8 antennas are needed. Since the size of a single antenna is increased to 16 mm, the gain is also increased to 15 dB, and the total gain of the array is still 24 dB. However, the system design can be substantially simplified due to the reduced number of antennas.

Due to the sparse array structure, the traditional angle estimation method cannot be directly applied. For a specific frequency component in the signal, the spatial spectrum of the sparse array structure will have multiple peaks at the same time, and the angle estimation is fuzzy. The traditional narrowband array cannot be employed to solve this problem. However, the wideband signal provides an idea for solving the AoA ambiguity problem. Considering the actual physical model of signal propagation, each frequency component in the wideband signal is incident on the receiving antenna array from the same angle, and therefore the incident angles corresponding to the respective frequency components are the same. At the same time, we carefully observe the corresponding response of the signal incident to the antenna array, where

$$e^{j2\pi[(f_k+f_c)d\frac{\sin\theta}{c}]m}. \tag{5.45}$$

We can find out that the phase of the signal is not only related to the angle of incidence θ, but also to the specific frequency component f_k. The array signal periodogram is a two-dimensional function of the incident angle θ and the frequency component f_k. Moreover, although for each frequency component f_k, the incident angle position of the periodic diagram reaction is ambiguous, that is, there are a plurality of spikes. However, considering all frequency components $f_0 \sim f_{K-1}$, only the position corresponding to the true incident angle will peak in the periodogram of all frequency components $f_0 \sim f_{K-1}$. Therefore, we can first construct a fuzzy-free AoA estimation method, which selects the incident angle with peaks on each frequency component as the angle estimate.

As shown in Fig. 5.3, a wideband signal from 260 to 340 GHz is incident on the antenna array from 30° Due to the extremely sparse spacing of the antennas, there are tens of peaks for any one frequency component, that is, dozens of false angle estimates, as shown in Fig. 5.3. However, only one true angle estimate exists in all frequency components. Thus, combining the spatial spectra of all frequency components, we obtain a single spatial spectrum that is independent of the frequency components. There is only one spectral line in this spatial spectrum, which corresponds to the true angle. Based on this result, we can find that even if a sparse array is used, the blur of the spatial spectrum estimation can be eliminated by refining the spectral structure of the signal.

In the following, we will demonstrate that the signals incident from different angles are approximately orthogonal. That is, in the broadband sparse array, there

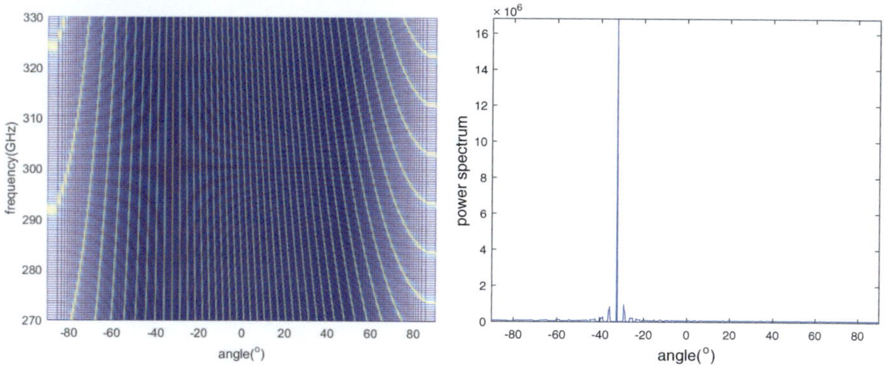

(a) Power spectrum with different frequency (b) Combined power spectrum

Fig. 5.3 Proof of orthogonality in a sparse array. Even though the separation between antennas is much larger than half-wave, by combining all the spatial spectra of all frequency components, only one peak occurs at the true angle. **a** For each frequency, there exist multiple peaks. **b** By combining, only one peak is left

is almost no angle estimation blur problem. In order to simplify the calculation, we first define $\phi = \sin\theta$, $|H_k X_k|^2 = \sigma_s^2$, $f_k = k\Delta f$, and $d = Q\frac{c}{f_c}$.

Therefore, we separately observe the two signals corresponding to the incident angle ϕ and the incident angle $(\phi + \Delta\phi)$ and calculate the inner product of these two terms.

$$\begin{aligned} Tr\left(S(\phi)^H S(\phi+\Delta\phi)\right) &= \sum_k \sum_m |H_k X_k|^2 e^{-j2\pi(f_k+f_c)m\frac{d\Delta\phi}{c}} \\ &= \sum_k \sum_m \sigma_s^2 e^{-j2\pi\left(1+\frac{k\Delta f}{f_c}\right)Q\Delta\phi m} \\ &= \sum_k \sigma_s^2 g(k), \end{aligned} \tag{5.46}$$

in which,

$$g(k) = e^{-j2\pi\left(1+\frac{k\Delta f}{f_c}\right)Q\Delta\phi\frac{M-1}{2}} \left[\frac{\sin\left(\pi\left(1+\frac{k\Delta f}{f_c}\right)Q\Delta\phi M\right)}{\sin\left(\pi\left(1+\frac{k\Delta f}{f_c}\right)Q\Delta\phi\right)}\right]. \tag{5.47}$$

For different $\Delta\phi$, we can have the following conclusion:

- When $\left(1+\frac{k\Delta f}{f_c}\right)Q\Delta\phi M \le \frac{1}{2}$, which means $\Delta\phi \le \frac{1}{QM}\frac{1}{\left(1+\frac{k\Delta f}{f_c}\right)} \approx \frac{1}{2QM}$, we have $\sin\left(\pi\left(1+\frac{k\Delta f}{f_c}\right)Q\Delta\phi\right) \approx \pi\left(1+\frac{k\Delta f}{f_c}\right)Q\Delta\phi$. We use integration to replace summation

$$\sum_k g(k) \approx Me^{-j2\pi Q\Delta\phi\frac{M-1}{2}} \sum_k e^{-j2\pi\left(\frac{k\Delta f}{f_c}\right)Q\Delta\phi\frac{M-1}{2}} sinc\left(\left(1+\frac{k\Delta f}{f_c}\right)Q\Delta\phi M\right)$$
$$\approx Me^{-j2\pi Q\Delta\phi\frac{M-1}{2}} \frac{f_c}{\Delta f}\int_{f_0/f_c}^{f_{K-1}/f_c} e^{-j2\pi Q\Delta\phi\frac{M-1}{2}x} sinc((1+x)Q\Delta\phi M)dx. \tag{5.48}$$

Considering that $f_k/f_c \ll 1$, the variable $x \approx 0$, leads to $sinc((1+x)Q\Delta\phi M) \approx sinc(Q\Delta\phi M)$ and $e^{-j2\pi Q\Delta\phi\frac{M-1}{2}} \approx 1$. Thus, we have

$$\sum_k g(k) \approx MKe^{-j2\pi Q\Delta\phi\frac{M-1}{2}} sinc(Q\Delta\phi M). \tag{5.49}$$

- When $\frac{1}{2QM} \le \Delta\phi \le \frac{1}{2Q}$, $(1+x)Q\Delta\phi M \ge \frac{1}{2}$, leads to $sinc((1+x)Q\Delta\phi M) \ll 1$, and $\sum_k g(k) \approx 0$. This is easy to understand: consistent with traditional narrow array, signals with angles separation over Rayleigh limit can be separated easily. Once the angle moves, signal power attenuates according to the sinc function.
- When $\frac{1}{2Q} \le \Delta\phi$, this is what we interested in. For traditional narrow arrays, antennas deployed with separation over half wavelength lead to a serious problem: Signal power occurs on more than one angle, which makes angle estimation impossible. However, in light of the different frequency components of the wideband signal, we can still find the true angle.

Recalling the previous discussion of the relationship between array size and symbol rate in large-scale arrays, we will find that the array size is much larger than the corresponding spatial distance in a symbol's space length, which means $(M-1)\lambda/2 \ge CT_f$. By a little mathematical manipulation, we find that $K\Delta f \ge 2f_c/(M-1)$, which also means $\frac{k\Delta f}{f_c} \ge \frac{2k}{K(M-1)}$.

Consider the inner product

$$\sum_k \sigma_s^2 g(k) = \sum_k e^{-j2\pi\left(1+\frac{k\Delta f}{f_c}\right)Q\Delta\phi\frac{M-1}{2}} \left[\frac{\sin\left(\pi\left(1+\frac{k\Delta f}{f_c}\right)Q\Delta\phi M\right)}{\sin\left(\pi\left(1+\frac{k\Delta f}{f_c}\right)Q\Delta\phi\right)}\right]$$
$$= \sigma_s^2 e^{-j2\pi Q\Delta\phi\frac{M-1}{2}} \sum_k e^{-j2\pi\left(\frac{k\Delta f}{f_c}\right)Q\Delta\phi\frac{M-1}{2}} \left[\frac{\sin\left(\pi\left(1+\frac{k\Delta f}{f_c}\right)Q\Delta\phi M\right)}{\sin\left(\pi\left(1+\frac{k\Delta f}{f_c}\right)Q\Delta\phi\right)}\right]. \tag{5.50}$$

Just take those fake angles into account, corresponding to $\Delta\phi = q/Q$, we can infer that (for these angles) in a traditional narrow array, there holds $\frac{\sin\left(\pi\left(1+\frac{k\Delta f}{f_c}\right)Q\Delta\phi M\right)}{\sin\left(\pi\left(1+\frac{k\Delta f}{f_c}\right)Q\Delta\phi\right)} \approx \frac{\sin(\pi Q\Delta\phi M)}{\sin(\pi Q\Delta\phi)} = const$. We thus simplify the previous summation as

$$\sum_k \sigma_s^2 g(k) = C\sum_k e^{-j2\pi\left(\frac{k\Delta f}{f_c}\right)Q\Delta\phi\frac{M-1}{2}}. \tag{5.51}$$

Here, we find the equivalent digital frequency satisfies $\left(\frac{k\Delta f}{f_c}\right) Q\Delta\phi\frac{M-1}{2} \geq q/K$, implying that this summation contains more than q periods sine wave and $\sum_k \sigma_s^2 g(k) \approx 0$.

Here, it should be emphasized that compared with the proof in Sect. 5.3.2, the proof in Sect. 5.4.1 focuses on solving the problem that in traditional sparse arrays multiple peaks may occur in angle spectrum, which does not exist in the proof in Sect. 5.3.2. The core of the proof in Sect. 5.4.1 is to explain that due to the increase of the signal bandwidth, at the angles that fake peaks should appear, due to the additional degrees of freedom provided by the wideband signal, there are more than one periodic sine wave summation in the integration result.

5.4.2 *Wideband Modelation Converter Structure*

Since the general modulation signal bandwidth can reach tens of GHz, it is far beyond the sampling rate of common ADC, even if the Nyquist sampling rate is satisfied. The extremely high sampling rate leads to huge signal processing complexity. Meanwhile, even though a sparse array is used, the number of antennas may be still large. Furthermore, the cost of RF chains could be unacceptable. Therefore, we need to further reduce the signal dimension. The signal sampling rate is reduced here by using a method called WMC, in which the RF chains are reduced in a similar way [26]. As shown in Fig. 5.4, the M received signals are first sent to a random analog matrix and combined into $M^{'}$ signals. Then, the $M^{'}$ signals are mixed with the locally generated periodic pseudo-random sequence square wave signal $p_{m'}(t)$, and then filtered by a low-pass filter, followed by sampled by a low-speed ADC to obtain a sampling sequence $S_{m'}(n)$.

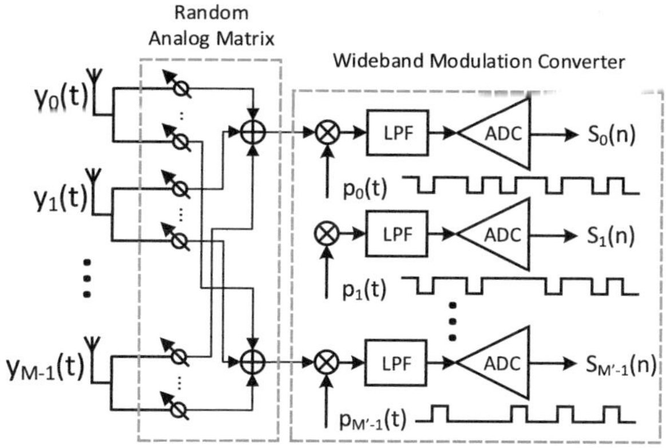

Fig. 5.4 2-D WMC structure. Signals ($y_m(t)$, $m = 0, 1, \ldots M-1$) received in M antennas are first combined into $M^{'}$ chains by random-analog-matrix. $M^{'}$ wideband signals are then converted into $M^{'}$ narrowband sampled signals ($S_{m'}(n)$, $m^{'} = 0, 1, \ldots M^{'}-1$) by wideband-modulation-converter

First, we deduce the signal after the random analog matrix as

$$Y'_{m',k} = \sum_{m} Y_{m,k}\beta_{m',m}, \tag{5.52}$$

where $\beta_{m',m}$ represents the random phase shifter between mth antenna and m'th RF chain. The random analog matrix reduces signal dimension by a factor M'/M.

We then deduce the signal after the WMC structure. The spectrum of the m'th pseudo-random sequence is expressed as follows. Since a pseudo-random sequence with periodic repetition is usually used, the spectrum of the signal can be written in the form of a number of series.

$$P_{m'}(f) = \sum_{k} c_{m',k}\delta(f - k\Delta f), \tag{5.53}$$

in which, Δf represents the repetition frequency of the pseudo-random sequence.

Thus, the m'th RF chain received signal $S_{m'}(f)$ after mixing can be written as

$$S_{m'}(f) = Y'_{m'}(f) * P_{m'}(f). \tag{5.54}$$

The signal $S_{m'}(f)$ after mixing is low-pass filtered. Since the THz signal bandwidth is large, usually on the order of tens of GHz, a low-pass filter with a very narrow bandwidth is used here to reduce the ADC sampling rate. The bandwidth of the filter can reach one percent of the signal bandwidth. Therefore, compared with other spectral lines, this chapter only considers the baseband signal component.

$$S_{m',k'} = \sum_{k} Y'_{m'}(k)P_{m'}(f_{k'-k}) = \sum_{k} c^*_{m',k'-k}Y'_{m',k} = \sum_{m}\sum_{k} Y_{m,k}\beta_{m',m}c^*_{m',k'-k}. \tag{5.55}$$

In realization, we can just use the same pseudo-random sequence, which means $c_{m',k'-k} = c_{k',k}$.

Thereby, we have the expression of $S_{m',k'}$ as

$$S_{m',k'} = \sum_{m}\sum_{k} \beta_{m',m}Y_{m,k}c_{k,k'} \implies [\boldsymbol{S}]_{M',K'} = [\boldsymbol{B}]_{M',M}[\boldsymbol{Y}]_{M,K}[\boldsymbol{C}]_{K,K'}. \tag{5.56}$$

In this way, we can get the ADC sampling signal as

$$\boldsymbol{s} = \boldsymbol{T}\boldsymbol{z} + \boldsymbol{n} = \boldsymbol{q} + \boldsymbol{n}. \tag{5.57}$$

Here, z is a long vector with KQM elements, in which each element represents whether there exists a signal in this grid. If signal comes from the jth grid, $z(j) = 1$, else $z(j) = 0$. At the same time, the jth column of $\boldsymbol{T}$ represents the response of the signal from jth gird. Thus, we give out the spare recovery problem, in which $\boldsymbol{T}$ is a flat observation matrix and z represents the angle of incidence of the signal in space. Therefore, we need to recover the sparse high-dimensional signal z from the low-dimensional observation signal $\boldsymbol{s}$. Through the random analog matrix, M signals are combined into $M^{'}$ signals. Furthermore, through the WMC structure, the original signal with a K-line is converted into a signal with only $K^{'}$ frequency lines. In practical scenarios, $M^{'} \ll \frac{M}{10}$ and $K^{'} \ll \frac{K}{10}$. The signal dimension can thus be reduced by two orders of magnitude.

5.4.3 GAMP-Based Joint Angle Delay Estimation

For the sparse signal recovery problem, it can be solved by orthogonal matching pursuit (OMP) and other methods. However, there are certain limitations in THz large-scale antenna systems. In order to make full use of the directional characteristics of the THz array and simplify the system design, the communication system usually assigns one RF chain to each user. Therefore, the number of users is usually only slightly less than the number of RF chains. In the case of less observation, even if the signal has spatial sparsity, the traditional OMP and other algorithms still can't work normally. Due to the small number of observations, the OMP and other algorithms will have matrix rank loss in the calculation, resulting in matrix inversion (i.e., the operation value becomes unstable). Therefore, this chapter uses the GAMP algorithm to solve sparse signal recovery.

First, we split the entire search space into J grids that do not overlap. Considering that the THz beam is extremely narrow and the delay is extremely short, each user occupies only one of the J discrete grids. When the system serves U users, for each beam delay block θ_j, whether the user exists on the beam delay block obeys the Bernoulli distribution.

Definition: the probability that there is no user is λ, then λ satisfies the condition of

$$\lambda = 1 - \frac{U}{J}. \tag{5.58}$$

Thus, the distribution of z can be written as

$$p(z) = \lambda\delta(z) + (1 - \lambda)\delta(z - 1). \tag{5.59}$$

In the GAMP algorithm, in order to simplify the calculation, we use the central limit theorem to approximate the Gaussian distribution of the intermediate variables $z, \boldsymbol{q}$ under the premise of large signal dimension. Thus, the conditional posterior distribution of $z, \boldsymbol{q}$ can be written as

$$
\begin{aligned}
P_{Z|S}(z|s;\hat{r},\mu^r) &= \frac{P_Z(z)\times N(z;\hat{r},\mu^r)}{\int_{z'} P_Z(z')\times N(\hat{z};\hat{r},\mu^r)} \\
&= \frac{[\lambda\delta(z)\times N(0;\hat{r},\mu^r)+(1-\lambda)\delta(z-1)N(1;\hat{r},\mu^r)]}{\lambda N(0;\hat{r},\mu^r)+(1-\lambda)N(1;\hat{r},\mu^r)}, \\
P_{Q|S}(q|s;\hat{q},\mu^q) &= \frac{P_{S|Q}(s|q)\times N(q;\hat{q},\mu^q)}{\int_{q'} P_{S|Q}(s|q')\times N(q';\hat{q},\mu^q)} \\
&= \frac{1}{C}N(s;q,\sigma_n^2)\times N(q;\hat{q},\mu^q) \\
&= N\left(q,\frac{s}{\sigma_n^2}+\frac{\hat{q}}{\mu^q},\frac{\mu^q\sigma_n^2}{\mu^q+\sigma_n^2}\right).
\end{aligned} \tag{5.60}
$$

Due to the Gaussian approximation, we can easily calculate the expectation and variance under the conditional posterior distribution of z.

$$
\begin{aligned}
E[Z|s;\hat{r},\mu^r] &= \int_z zP_{Z|Y}(z|s;\hat{r},\mu^r) \\
&= P_{Z|Y}(1|s;\hat{r},\mu^r) = \left[1+\frac{\lambda}{1-\lambda}\times\frac{N(0;\hat{r},\mu^r)}{N(1;\hat{r},\mu^r)}\right]^{-1}, \\
VAR[Z|s;\hat{r},\mu^r] &= \int_z |z-E[Z|s;\hat{r},\mu^r]|^2 P_{Z|Y}(z|s;\hat{r},\mu^r) \\
&= E[Z|s;\hat{r},\mu^r]-E[Z|s;\hat{r},\mu^r]^2 \\
&= \left(\frac{\lambda}{1-\lambda}\times\frac{N(0;\hat{r},\mu^r)}{1;\hat{r},\mu^r}\right)\left(1+\frac{\lambda}{1-\lambda}\times\frac{N(0;\hat{r},\mu^r)}{N(1;\hat{r},\mu^r)}\right)^{-2}.
\end{aligned} \tag{5.61}
$$

Meanwhile, because $P_{Q|S}(q|s;\hat{q},\mu^q)$ is a product of two Gaussian distributions, we can easily calculate the expectation and variance of $\boldsymbol{q}$.

$$
\begin{aligned}
E_{Q|S}[q|s;\hat{q},\mu^q] &= \frac{s}{\sigma_n^2}+\frac{\hat{q}}{\mu^q}, \\
VAR_{Q|S}[q|s;\hat{q},\mu^q] &= \frac{\mu^q\sigma_n^2}{\mu^q+\sigma_n^2}.
\end{aligned} \tag{5.62}
$$

We further define the following functions.

$$
\begin{aligned}
g_{in}(\hat{r},\mu^r) &= E[Z|s;\hat{r},\mu^r] = \left[1+\frac{\lambda}{1-\lambda}\times\frac{N(0;\hat{r},\mu^r)}{N(1;\hat{r},\mu^r)}\right]^{-1},\\
g_{in}^{'}(\hat{r},\mu^r) &= \frac{1}{\mu^r}VAR[Z|s;\hat{r},\mu^r] = \frac{1}{\mu^q}[g_{in}(\hat{r},\mu^r)-g_{in}(\hat{r},\mu^r)^2],\\
g_{out}(s,\hat{q},\mu^q) &= \frac{1}{\mu^q}(E_{Q|S}[q|s;\hat{q},\mu^q]-\hat{q}) = \frac{s-\hat{q}}{\mu^q+\sigma_n^2},\\
g_{out}^{'}(s,\hat{q},\mu^q) &= \frac{1}{\mu^q}\left(\frac{VAR_{Q|S}[q|s;\hat{q},\mu^q]}{\mu^q}-1\right) = \frac{-1}{\mu^q+\sigma_n^2}.
\end{aligned}
\tag{5.63}
$$

The algorithm is described as Algorithm 20.

Algorithm 5.1 GAMP-based joint angle-delay estimation algorithm

for each $n \in [0, N-1]$ **do**
  initialize $\hat{z}_n(1) = E(z) = (1-\lambda)$;
  initialize $\mu_n^z(1) = VAR(z) = (1-\lambda)\lambda$;
end for
for each $m \in [0, M-1]$ **do**
  initialize $\hat{u}_m(0) = 0$;
end for
while $t \leq Iter_{max}$ **do**
  for each $m \in [0, M-1]$ **do**
    $\mu_m^p(t) = \sum_{n=0}^{N-1} |B_{m,n}|^2 \mu_n^z(t)$;
    $\hat{p}_m(t) = \sum_{n=0}^{N-1} B_{m,n}\hat{z}_N(t) - \mu_m^p(t)\hat{u}_m(t-1)$;
    $\hat{u}_m(t) = g_{out}(s_m, \hat{p}_m(t), \mu_m^p(t))$;
    $\mu_m^u(t) = -g_{out}^{'}(s_m, \hat{p}_m(t), \mu_m^p(t))$;
  end for
  for each $n \in [0, N-1]$ **do**
    $\mu_n^r(t) = (\sum_{m=0}^{M-1} |B_{m,n}|^2 \mu_m^u(t))^{-1}$;
    $\hat{r}_n(t) = \hat{z}_n(t) + \mu_n^r(t) \sum_{m=0}^{M-1} B_{m,n}^* \hat{u}_m(t)$;
    $\mu_n^z(t+1) = \mu_n^r g_{in}^{'}(\hat{r}_n(t), \mu_n^r(t))$;
    $\hat{z}_n(t+1) = g_{in}(\hat{r}_n(t), \mu_n^r(t))$;
  end for
end while

For z that needs to be estimated, the GAMP algorithm produces a list of estimates $z_n(t)$ and its corresponding variance $\mu_n^z(t)$.

5.5 Numerical Results and Analysis

5.5.1 Parameters Description

The spacing of the antenna array is $Q\lambda$. There are a total of M antennas, and the corresponding overall array size is $MQ\lambda = D\lambda$. The number of symbols of the pilot signal is I. According to the Nyquist sampling theorem, two sample points are sampled on one symbol, and the corresponding normalized frequency is $K = 2I$. Since the overall size of the antenna array is $MQ\lambda$, the corresponding Rayleigh limit of the angle estimation is $1/MQ$, and the angle estimation range is $[-0.5, 0.5]$. Therefore, the space required for the overall search is MQ. Since the signal normalized frequency range is $[-0.5/K, +0.5/K]$, the corresponding delay estimation Rayleigh limit is $1/K$, implying that the delay search space is K.

5.5.2 *Estimation Under the Rayleigh Limit*

1. Traditional MLE under half-wave array

In the traditional large-scale half-wave spacing antenna array system, in order to achieve the Rayleigh limit estimation accuracy, the dimension of the received signal is $M \times K$, and the corresponding overall search space is $MQ \times K$. Each search includes an integral conjugate multiplication of the primary signal with the local template, an overall summation, and a modulo operation. Therefore, the calculation amount of one search is $(MK + 1)$ complex multiplications and $(MK - 1)$ additions, we only consider the computational complexity of multiplications. Therefore, for the $MQ \times K$ search, the overall computational complexity is $O(M^2K^2Q)$.

2. FFT-based MLE on half-wave array

First, K times MQ point FFT is calculated for the received MK-dimensional signal, and after each FFT, the MQ point result needs to be interpolated in turn. Since there are only M observation signals for each frequency component, the actual MQ point FFT is obtained by M-point FFT cycle extension. In practice, the computational complexity of a FFT is only $M\log(M)$. Each interpolation comprises 4 multiplications and 3 additions, wherein the interpolation coefficients can be stored in advance and do not need to be separately calculated. Therefore, the computational complexity of the K-order MQ point correction FFT is $(O(KM\log(M)) + O(4KMQ))$. Then, the second time delay estimation is performed, and the MQ-order K-point conjugate multiplication is required, and the corresponding computational complexity is $O(KMQ)$. Perform MQ minimum K-point IFFT operation on the calculation result, and the corresponding computational complexity is $O(MQK\log(K))$. Finally, the KM results are subjected to modulo operations, corresponding to KM multiplication operations. Therefore, the overall computational complexity is $(O(KDlog(M)) + O(KDlog(K)))$.

3. MLE in sparse array

Sparse arrays have the same computational complexity representation as traditional arrays. Under the traditional MLE, the computational complexity of the sparse array is $O(M^2K^2Q) = O(MK^2D)$. Under FFT-MLE, the computational complexity of the sparse array is $(O(KMlog(M)) + O(KMQlog(K))) = (O(KD/Qlog(M)) + O(KDlog(K)))$. However, since a sparse array is employed, the number of M is reduced in the case where the antenna size $QM = D$ is fixed. Therefore, by comparison, we can find that under both algorithms, the computational complexity of sparse arrays is greatly reduced.

4. GAMP computational complexity under sparse array WMC
Under the sparse array WMC, compared with the traditional array, on the one hand, the size of the antenna array is unchanged and the number of antennas is greatly reduced; on the other hand, the space-frequency WMC structure can greatly reduce the signal bandwidth to be reduced to $K^{'}$. Also, reduce the number of RF channels to $M^{'}$. Thus, the dimension of the observed signal is $M^{'}K^{'}$. The space that needs to be estimated is still MQK. Thus, the dimension of the observation matrix is $M^{'}K^{'} \times MQK$. In each iteration, it is critical to calculate the multiplication of the cubic matrix and the vector. The corresponding computational complexity is $O(M^{'}K^{'} \times MQK)$, and the rest is mainly the vector multiplied by the elements, the computational complexity is respectively $O(M^{'}K^{'})$ and $O(MQK)$. Therefore, considering that a total of N iterations have been performed, the corresponding computational complexity is $O(NM^{'}K^{'} \times MQK)$. Since in practice $M^{'} \ll M, K^{'} \ll K$, N is a constant not exceeding 10, the GAMP-based algorithm can greatly reduce the computational complexity.
Table 5.1 shows the complexity of the above four algorithms.

Table 5.1 Complexity of estimation under Rayleigh limit

Estimator	Complexity
MLE under half-wave array	$O(M^2K^2Q)$
FFT-based MLE	$O(KD\log(M)) + O(KDlog(K))$
MLE in sparse array	$O(KD/Q\log(M)) + O(KD\log(K))$
GAMP-based estimator	$O(NM^{'}K^{'} \times MQK)$

5.5.3 Progressively Reaching CRLB Estimation

1. Traditional MLE method
For the progressively reaching CRLB estimation algorithm, the traditional MLE-based estimation method is mainly realized by increasing the grid precision. Therefore, the computational complexity of progressive CRLB is a multiple of the computational complexity based on the Rayleigh limit. Since we already know CRLB, we can indirectly get the meshing accuracy of the mesh through CRLB. Here, for the convenience of calculation, we directly give out the magnitude of CRLB:

$$
\begin{aligned}
var(\theta) \geq I(\theta,\tau)^{-1}_{\theta,\theta} &\approx \frac{1}{\sum_{k=0}^{K-1}|H_k X_k|^2\left(2\pi\left(\frac{f_k}{f_c}\right)+1\right)^2}\frac{12}{M(M^2-1)}(Q\cos\theta)^{-2} \\
&= \frac{1}{Z_\theta^2}\frac{1}{KM^3Q^2}, \\
var(\tau) \geq I(\theta,\tau)^{-1}_{\tau,\tau} &\approx \frac{1}{(2\pi)^2\sum_{k=0}^{K-1}|H_k X_k|^2 k^2} \\
&= \frac{1}{Z_\tau^2}\frac{1}{K^3M}.
\end{aligned}
\tag{5.64}
$$

Among them, Z_θ and Z_τ are much larger than 1 at high SNR. Thus, at high SNR, the accuracy of the grid should be at least higher than the standard deviation of the CRLB. For angle estimation, the Rayleigh limit is $1/MQ$, implying that the search accuracy corresponding to CRLB is $1/\sqrt{Z_\theta^2 KM^3Q^2}$, and the computational complexity of the progressive estimation needs to increase at least $Z_\theta\sqrt{KM}$ times. For the delay estimation, the Rayleigh limit is $1/K$, the search accuracy corresponding to CRLB is $1/\sqrt{Z_\tau^2 K^3 M}$, and the computational complexity of the progressive estimation needs to increase at least $Z_\tau\sqrt{KM}$ times. Therefore, the computational complexity overall needs to increase $Z_\theta Z_\tau MK$ times, corresponding to $O(Z_\theta Z_\tau M^3K^3Q) = O(Z_\theta Z_\tau M^3K^2D)$.

2. FFT-based MLE method
Similar to pure MLE, in order to improve the accuracy, it is necessary to increase the number of points of the FFT twice to $Z_\theta\sqrt{KM}$ times and $Z_\tau\sqrt{KM}$ times. Therefore, the total computational complexity is $(O(Z_\theta K^{1.5}M^{1.5}\log(Z_\theta M^{1.5}K^{0.5})) + O(Z_\tau Z_\theta K^2M^2\log(Z_\tau KM^{0.5})))$.

3. Calculation complexity based on phase combination estimation
The phase combination-based estimation method is based on the Rayleigh limit estimation and the local phase estimation is added. Therefore, the computational complexity is only relative to the Rayleigh limit, which increases the complexity of the phase estimation. For the arctan function, it is actually implemented using one division operation and a fifth-order approximation function. Therefore, it is first necessary to perform phase correction on the KM-dimensional signal, corresponding to KM complex multiplication, and then perform phase estimation for each element, which requires 6 KM multiplication. A total of 7 KM multiplications are required. Taking the FFT-based method as an example, the computational complexity is $(O(KM\log(M)) + O(KM\log(K)))$.
Table 5.2 shows the complexity of the above three algorithms.

Table 5.2 Complexity of estimation progressively reaching CRLB

Estimator	Complexity
MLE with fine grid	$O(Z_\theta Z_\tau M^2 K^2 D)$
FFT-based MLE	$O(Z_\theta K^{1.5} M^{1.5} \log(Z_\theta M^{1.5} K^{0.5}))$ $+O(Z_\tau Z_\theta K^2 M^2 \log(Z_\tau K M^{0.5}))$
Phase combination estimator	$O(KM \log(M)) + O(KM \log(K))$

5.5.4 Numerical Simulation

This section verifies the algorithm described by numerical simulation. First, we verified the feasibility of multiuser joint angle and delay estimation under half-wave array conditions. The performances of the MLE estimation method and the linear phase combination estimation method are evaluated separately and compared with the multiple signal classification (MUSIC) algorithm, ROOT-MUSIC algorithm, and CRLB. Then, we verified that the algorithm proposed in this chapter can still work under the sparse array. Similarly, the result is compared with the MUSIC algorithm, ROOT-MUSIC algorithm, and CRLB. Finally, we present the estimation results based on the GAMP algorithm. By using the WMC structure, it can support simultaneous estimation of a large number of users with only a small number of antennas and RF chains. Moreover, the MUSIC method and root-MUISC method are adopted as two comparisons [36–38].

Joint angle and delay estimation simulation is performed on a large-scale antenna array with a traditional half-wave interval. A large-scale array of 256 antennas is used, the antenna spacing is 0.5λ, the signal carrier frequency is 300 GHz, and the BPSK modulation method with a symbol rate of 10 Gsps is adopted. The pilot uses a pseudo-random sequence of 8 symbols.

Figure 5.5 shows the space-time two-dimensional MLE estimation results in the presence of 10 users. Each peak in the graph corresponds to a user, the angle of incidence, and delay of the signal correspond to the coordinates of the peak. Obviously, the coordinate precision of the spike is limited by the accuracy of the meshing. Any point in the grid represents an inner product of the received signal and the local template. The calculation is KM multiplications and $(KM - 1)$ additions. Corresponding to the traditional Rayleigh limit, the number of spatial grids is M, and the number of time grids is K. Thus, a complete estimate requires a computational amount of $(KM)^2$. A more elaborate meshing means a multiple of the amount of computations.

The simulation is carried out for the case of a half-wave array, and the number of antenna elements is 32, 64, and 128, respectively. Due to the Rayleigh limit performance simulation, the accuracy of the angle of incidence grid is limited by the overall size of the array $(1/MQ)$, and the accuracy of the time delay grid is limited by the bandwidth of the signal $(1/B)$. Due to the meshing, the angle and delay estimates have a backplane effect in the case of a high SNR.

It can be seen from Fig. 5.6 that in the angle estimation and delay estimation, the MLE algorithm proposed in this chapter is better than the 2D-MUSIC algorithm. This is because 2D-MUSIC needs to estimate the angle first, and then estimate delay based on the angle estimation result. Due to the use of wideband signals, when MUSIC estimation is used, several spectral lines in the wideband signal need to

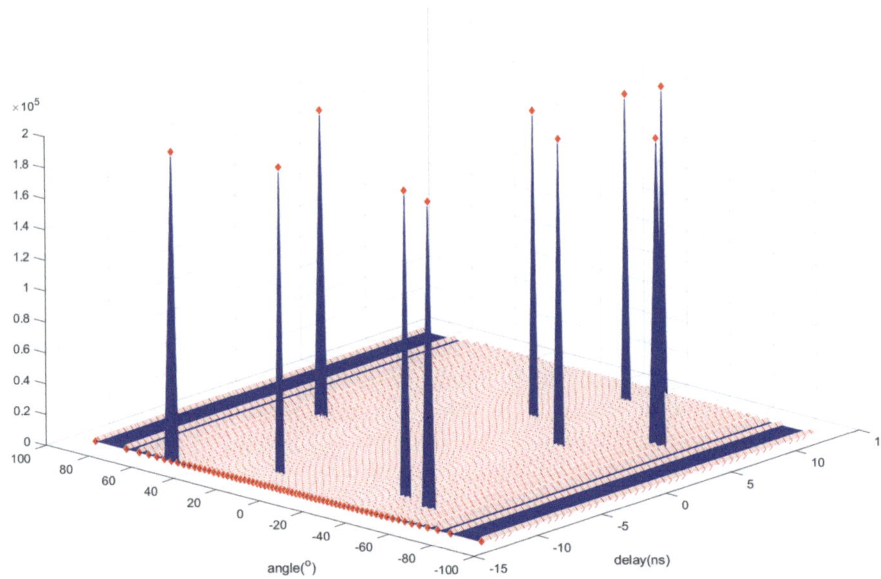

Fig. 5.5 Multiuser joint-angle-delay-estimation in large-scale half-wave array. We set $M = 256$, the center frequency is 300 GHz and the symbol rate is 10 Gsps

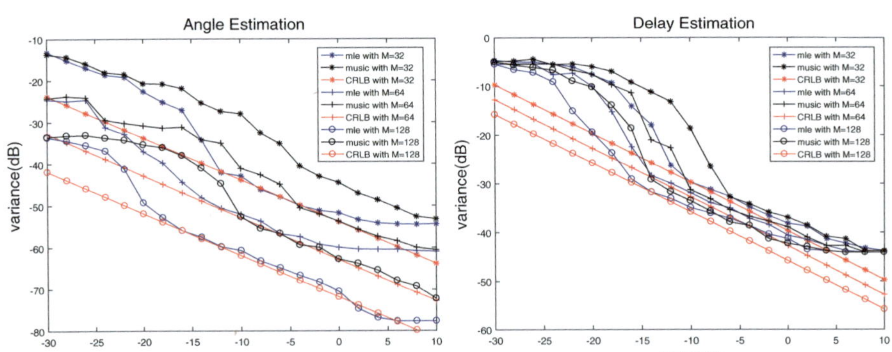

(a) Angle Estimation (b) Delay Estimation

Fig. 5.6 MLE versus 2D-MUSIC in half-wave array. Compare angle and delay estimation accuracy in the MLE method, 2D-MUSIC method, and CRLB. We set the center frequency as 300 GHz, symbol rate as 30 Gsps. **a** Angle estimation under 2 methods **b** Delay estimation under 2 methods

be independently estimated and then combined. Since there is no prior information for the delay estimate, the angle estimates for each line can only be combined in an independent form, which results in a loss of SNR. Also, since the estimation of delay depends on the result of angle estimation, the estimation accuracy of delay will also deteriorate. It can be seen from Fig. 5.6a and b that as the SNR increases, the estimation error of the MLE method quickly approaches CRLB, and the estimation error of the MUSIC method always has a certain gap with the CRLB, and there is a fixed interval between the two. The reason for this interval is that there is a loss of SNR when the MUSIC spatial spectrum is combined under different frequency components in the wideband signal. In addition, since the spatial spectrum of the MUSIC algorithm is the reciprocal form of the noise spatial spectrum, the estimation result of the MUSIC algorithm is susceptible to noise in the case of low SNR. Corresponding to the low SNR region, the MLE estimation method estimates better. The MLE estimation proposed in this chapter is a joint estimation of angle and delay directly, so there is no deterioration of SNR. Therefore, it is superior to the traditional MUSIC algorithm in two aspects.

Here, we explain in detail the floor effect in the simulation results. The backplane effect here is mainly due to meshing. 1. For angle estimation, the grid partitioning standard is limited by the Rayleigh limit of the array size. Therefore, as the M and Q increase, the array size also increases, and the corresponding angle estimates the Rayleigh limit according to $(1/MQ)$. The magnitude is decremented, and as the MQ increases, the backplane is also declining, but when the estimation accuracy is reduced to the $(1/MQ)$ level, it cannot be further improved. If you need to further improve the estimation accuracy, you need more fine meshing, which corresponds to the increase of computational complexity. 2. For TIME estimation, the division of the grid is limited by the bandwidth of the signal. Only increasing M and Q represents the increase of the overall size of the antenna array, and has no effect on the signal bandwidth. According to the traditional matched filtering ML scheme, the accuracy of the delay estimation is received. To the constraint of the ADC sampling rate (i.e., Nyquist sampling rate), the estimation accuracy is on the same order of magnitude as the sampling interval. 3. Under MLE estimation, the meshing accuracy is limited by the array size and signal bandwidth. Therefore, as long as the meshing accuracy is fixed, the estimation accuracy will have a backplane effect. In order to improve the estimation accuracy, the accuracy of meshing must be improved at the cost of computational complexity.

The simulation is carried out for the case of a half-wave array, and the number of antenna elements is 16, 32, and 64, respectively. The Fig. 5.7a and b shows the result. Due to the meshless estimation method, the estimation method provided in this chapter quickly reaches CRLB for angle estimation and delay estimation. In the root-MUSIC method, on the one hand, the non-coherent merging of the spatial spectrum of each frequency component in the wideband signal is required, resulting in the loss of SNR. On the other hand, in the case of a large number of array elements, there exists instability in the root-MUSIC. These two aspects degrade the root-MUSIC performance in the angle estimation. With the increase of the SNR, the estimation accuracy separation does not decrease with respect to the CRLB but increases.

In the delay estimation, the estimation accuracy of both methods approaches the CRLB. The approach speed of the two is basically the same. For the root-MUSIC algorithm, this is due to the fact that in the delay estimation, the pilot length is relatively short, the number of spectral lines in the corresponding wideband signal is relatively small, and the root-MUSIC algorithm is more accurate in finding the root result. In the time delay estimation, since the estimation of the angle has been completed, which is equivalent to having obtained the prior information on the angle, the performance of the root-MUSIC method for estimating the time delay is similar to that of the angle proposed in this chapter. The performance of the joint time-delay estimation method is similar.

Aiming at the problem of a large number of antennas in the traditional half-wave interval large-scale antenna array, the high-gain antenna sparse array structure is used to perform joint angle and delay estimation simulation. Adopting 16 antennas, the antenna spacing is 64 times λ, the signal carrier frequency is 300 GHz, and the BPSK modulation method with a symbol rate of 10 Gsps is adopted. The pilot uses a pseudo-random sequence of 8 symbols.

Figure 5.8 shows that even if the array spacing is much larger than the signal wavelength, accurate joint angle and delay estimation can still be achieved by the array's broadening effect on the signal. Moreover, the delay estimation accuracy can reach the Rayleigh limit of the signal bandwidth, and the accuracy of the incident angle estimation can also reach the Rayleigh limit of the overall size of the array.

For the case of a sparse array, the number of antenna elements is $M = 4$, 8, and 16, and the antenna spacing is Q = 8 times the wavelength. Due to the Rayleigh limit performance simulation, the accuracy of the angle of incidence grid is limited by the overall size of the array $(1/MQ)$, and the accuracy of the time delay grid is limited by the bandwidth of the signal $(1/B)$. In the same way as the half-wave array, due to

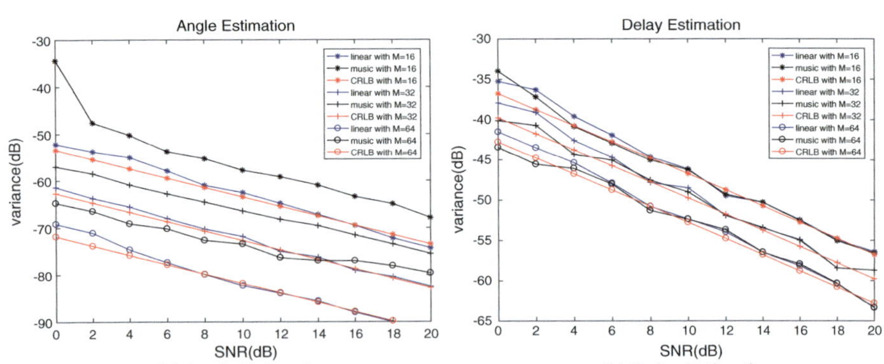

(a) Angle Estimation (b) Delay Estimation

Fig. 5.7 Linear phase combination off-grid estimation versus root-MUSIC estimation in the half-wave array. Compare angle and delay estimation accuracy in linear phase combination off-grid estimation, root-MUSIC estimation, and CRLB. We set the center frequency as 300 GHz and the symbol rate as 30 Gsps. **a** Angle estimation under 2 methods **b** Delay estimation under 2 methods

the mesh division, in the case of high SNR, the angle and delay estimates also have a backplane effect, and the final estimation error tends to the grid precision.

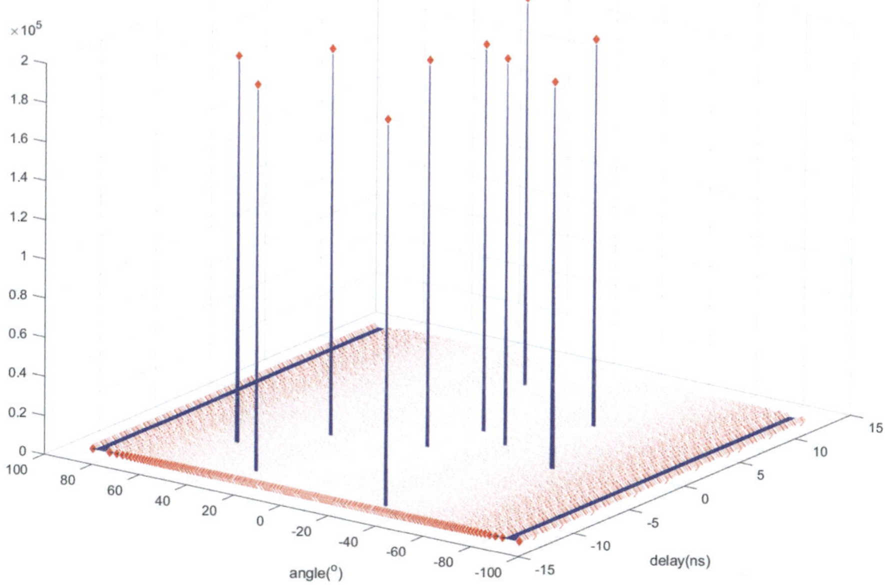

Fig. 5.8 Multiuser joint-angle-delay-estimation in sparse array. We set M = 16, the antenna separation as 64λ, the center frequency as 300 GHz, and the symbol rate as 10 Gsps

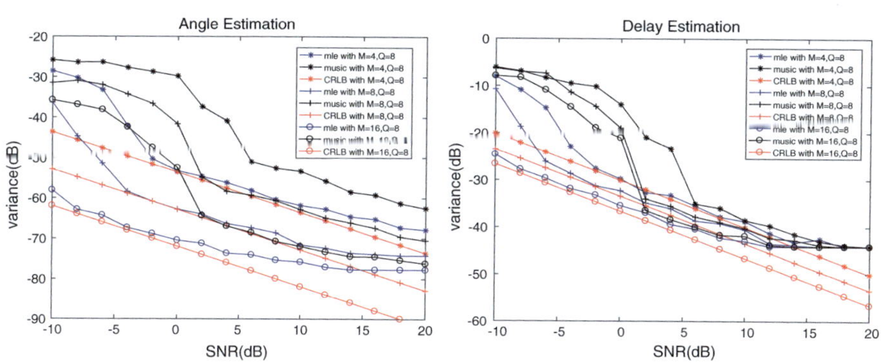

(a) Angle Estimation

(b) DelayEstimation

Fig. 5.9 MLE versus 2D-MUSIC in a sparse array. Compare angle and delay estimation accuracy in the MLE method, 2D-MUSIC method, and CRLB. We set the center frequency as 300 GHz and the symbol rate as 30 Gsps. **a** Angle estimation under 2 methods **b** Delay estimation under 2 methods

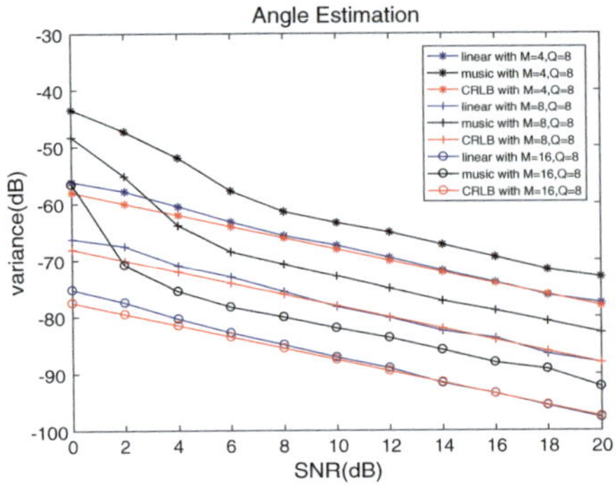

(a) Angle Estimation

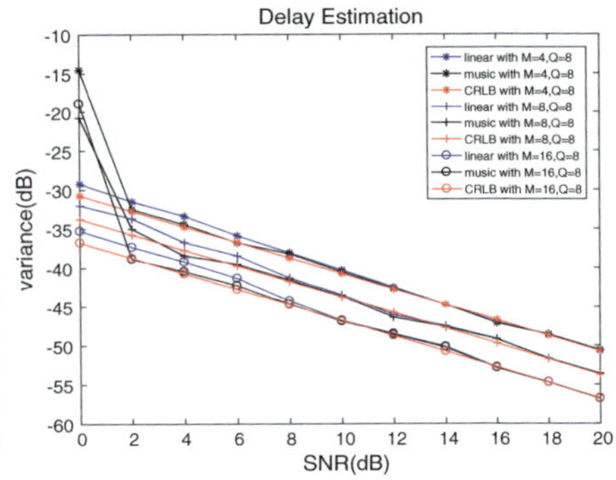

(b) Delay Estimation

Fig. 5.10 Linear phase combination off-grid estimation versus root-MUSIC estimation in a sparse array. Compare angle and delay estimation accuracy in linear phase combination off-grid estimation, root-MUSIC estimation, and CRLB. We set the center frequency as 300 GHz, and the symbol rate is 30 Gsps. **a** Angle estimation under 2 methods **b** Delay estimation under 2 methods

Through Fig. 5.9, we can find that the MLE algorithm proposed in this chapter is still better than the 2D-MUSIC algorithm under the condition of the sparse array. Although the estimation accuracy is limited by the meshing accuracy, the MLE algorithm can quickly converge to CRLB when the SNR is low, and the convergence speed of MUSIC is slower. For angle estimation, the MUSIC algorithm still has a delay to estimate the loss of spatial spectrum non-coherent combining under unknown conditions. In the delay estimation, although the MUSIC algorithm can also be CRLB within a certain range, it can be seen from Fig. 5.9 that the approach speed of MLE is much higher than that of the MUSIC algorithm. It is worth noting that comparing the sparse array with the traditional array, we choose the case of $M = 128$, $d = 0.5\lambda$ under the traditional array, and $M = 4$ and $d = 16\lambda$ under the sparse array. In both cases, the overall size of the array is 64. However, since M is three times and d is two in the denominator of CRLB, the influence of M is larger than d as a whole. Thus, the CRLB of the conventional array is still smaller than the sparse array of the same size.

For the simulation of the sparse array in Fig. 5.10, the number of antenna elements is M = 4, 8, and 16, and the antenna spacing is $Q = 8$ times the wavelength. Similar to the case of a half-wave array, the estimation method provided in this chapter quickly reaches CRLB. In the root-MUSIC method, on the one hand, the SNR loss is caused by the non-coherent merging of the spatial spectrum of each frequency component in the wideband signal, and on the other hand, the array is sparsely arranged relative to the half-wave array. The number of elements in the array is greatly reduced, which can improve the stability of the root result. Thus, the estimation accuracy of the root-MUSIC algorithm has a substantially fixed interval relative to the CRLB. In the delay estimation, the estimation accuracy of both methods approaches the CRLB. The approach speed of the two is basically the same.

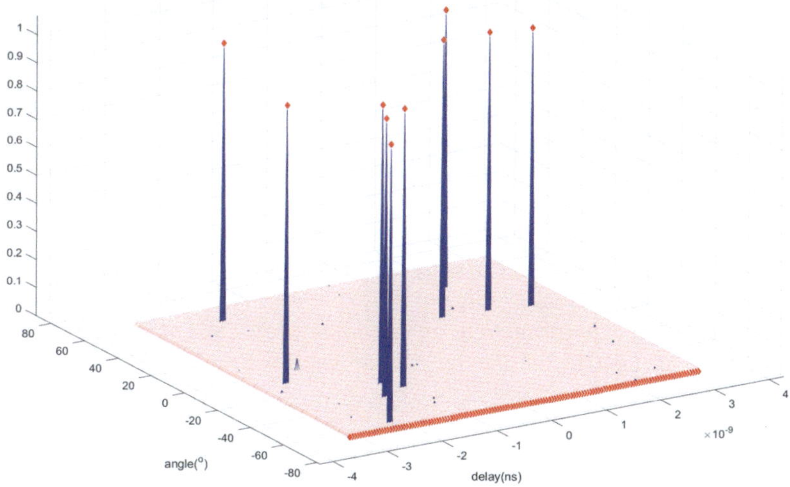

(a) Estimation with 10 users

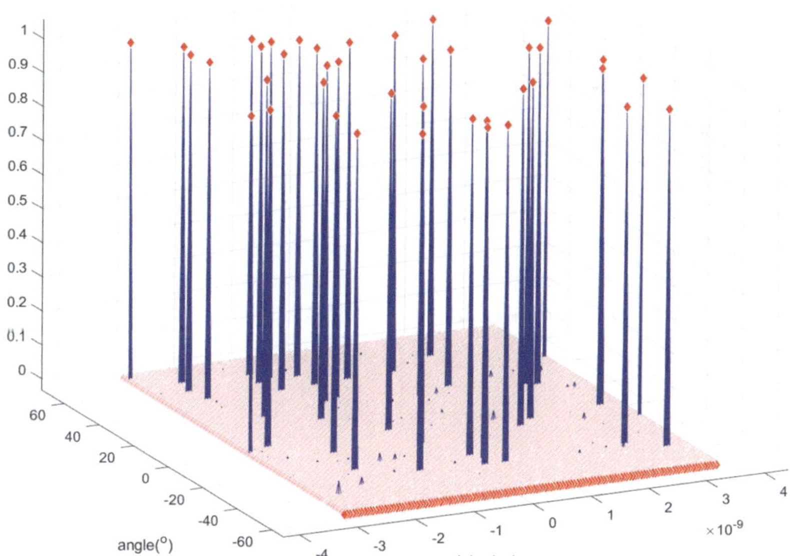

(b) Estimation with 40 users

Fig. 5.11 Multiuser joint-angle-delay estimation in sparse array with WMC structure and GAMP algorithm. **a** Estimation of 10 users by only 2 RF chians **b** Estimation of 40 users by only 16 antennas

For the case of multiple users, the joint-angle-delay-estimation of the 10 users sharing pilots is performed under the condition that the antenna interval is 32 times the wavelength, the number of antennas is 16 and the number of RF chains is only 2. The signal center frequency is 300 GHz and the symbol rate is 10 Gsps. Under the traditional array structure, this estimation cannot be achieved. However, by using the traditional CS framework, with large-scale sparse arrays, such high-precision estimation can be achieved through the GAMP algorithm. By using the WMC architecture, accurate estimates can be achieved using only 4.69% of the total bandwidth of the signal. Moreover, the estimation accuracy is consistent with the conventional array of the same size, and the Rayleigh limit can be obtained.

After simulation, we can see from Fig. 5.11a that the joint angle and delay estimation of 10 users can be realized by only using a large-scale array of 2 RF chains. Figure 5.11b shows that if 16 antennas are used, joint angle and delay estimation can be performed for up to 40 users, and the estimation error corresponds to the Rayleigh limit of the array size.

5.6 Conclusions

In this chapter, we investigate the joint estimation of angle and delay in a THz large-scale antenna array. Different from the traditional narrow-band assumption, in the large-scale THz array, we construct a new received signal model. Based on this model, a joint angle and delay estimation method based on MLE is given. In order to improve the estimation accuracy and avoid the huge computation complexity caused by fine grids, a progressive reaching CRLB estimation based on phase estimation is proposed. To deal with the problem of high dimension of signal in the large-scale THz array, we present a CS method based on a random phase matrix and WMC structure and solve this problem by GAMP algorithm. Simulation shows that, compared with the classic MUSIC algorithm, the proposed method has better performance. The GAMP method can simultaneously achieve multiuser parameter estimation with only a small number of antennas and RF chains.

References

1. Han C, Akyildiz IF, Gerstacker WH (2017) Timing acquisition and error analysis for pulse-based Terahertz band wireless systems. IEEE Trans Veh Technol 66(11):10102–10113
2. Liu Q, Wang Y, Fathy AE (2012) A compact integrated 100 Gs/s sampling module for UWB see through wall radar with fast refresh rate for dynamic real time imaging. In: IEEE Radio Wireless Symposium. IEEE, pp 59–62
3. Sarantos CH, Dagli N (2010) A photonic analog-to-digital converter based on an unbalanced mach-zehnder quantizer. Opt Express
4. Han C, Bicen AO, Akyildiz IF (2016) Multi-wideband waveform design for distance-adaptive wireless communications in the Terahertz band. IEEE Trans Signal Process 64(4):910–922

5. Wang B, Gao F, Jin S et al (2018) Spatial- and frequency-wideband effects in millimeter-wave massive MIMO systems. IEEE Trans Signal Process 66(13):3393–3406
6. Jian M, Gao F, Tian Z et al (2019) Angle-domain aided UL/DL channel estimation for wideband mmWave massive MIMO systems with beam squint. IEEE Trans Wireless Commun 18(7):3515–3527
7. Xin R, Ni Z, Wu S et al (2019) Low-complexity joint channel estimation and symbol detection for OFDMA systems. China Commun 16(7):49–60
8. Gul MMU, Ma X, Lee S (2015) Timing and frequency synchronization for OFDM downlink transmissions using Zadoff-Chu sequences. IEEE Trans Wireless Commun 14(3):1716–1729
9. Amor SB, Affes S, Bellili F et al (2019) Multi-node ML time and frequency synchronization for distributed MIMO-relay beamforming over time-varying flat-fading channels. IEEE Trans Commun 67(4):2702–2715
10. Zhang R, Xia W, Yan F et al (2019) A single-site positioning method based on TOA and DoA estimation using virtual stations in NLOS environment. China Commun 16(2):146–159
11. Feng R, Liu Y, Huang J et al (2018) Wireless channel parameter estimation algorithms: Recent advances and future challenges. China Commun 15(5):211–228
12. Shen X, Liao Y, Dai X et al (2018) Joint channel estimation and decoding design for 5G-enabled V2V channel. China Commun 15(7):39–46
13. Cheng F, Gui G, Zhao N et al (2019) UAV-relaying-assisted secure transmission with caching. IEEE Trans Commun 67(5):3140–3153
14. Zhao N, Lu W, Sheng M et al (2019) UAV-assisted emergency networks in disasters. IEEE Wireless Commun 26(1):45–51
15. Zhao N, Pang X, Li Z et al (2019) Joint trajectory and precoding optimization for UAV-assisted NOMA networks. IEEE Trans Commun 67(5):3723–3735
16. Gao Y, Xu J, Jia X (2015) Joint number and DoA estimation via the eigen-beam mCapon method for closely spaced sources. Sci China Inf Sci 58(12):217–219
17. Xiong X, Deng Z, Qi W et al (2019) High-precision angle estimation based on phase ambiguity resolution for high resolution radars. Sci China Inf Sci 62(4):53–55
18. Rong B, Zhang Z, Zhao X et al (2019) Robust superimposed training designs for MIMO relaying systems under general power constraints. IEEE Access 7(1):80404–80420
19. Ge Y, Zhang W, Gao F et al (2019) Angle-domain approach for parameter estimation in high-mobility OFDM with fully/partly calibrated massive ULA. IEEE Trans Wireless Commun 18(1):591–607
20. Stein Ioushua S, Yair O, Cohen D et al (2017) CaSCADE: Compressed carrier and DoA estimation. IEEE Trans Signal Process 65(10):2645–2658
21. Gupta P, Agrawal M (2019) Design and analysis of the sparse array for DoA estimation of noncircular signals. IEEE Trans Signal Process 67(2):460–473
22. Lin B, Liu J, Xie M et al (2015) Direction-of-arrival tracking via low-rank plus sparse matrix decomposition. IEEE Antennas Wireless Propag Lett 14:1302–1305
23. Qin G, Amin MG, Zhang YD (2019) DoA estimation exploiting sparse array motions. IEEE Trans Signal Process 67(11):3013–3027
24. Guo M, Zhang YD, Chen T (2018) DoA estimation using compressed sparse array. IEEE Trans Signal Process 66(15):4133–4146
25. Yang X, Sun Y, Zeng T et al (2016) Fast STAP method based on PAST with sparse constraint for airborne phased array radar. IEEE Trans Signal Process 64(17):4550–4561
26. Bar-Ilan O, Eldar YC (2014) Sub-Nyquist radar via doppler focusing. IEEE Trans Signal Process 62(7):1796–1811
27. Cohen D, Cohen D, Eldar YC et al (2018) SUMMeR: Sub-Nyquist MIMO radar. IEEE Trans Signal Process 66(16):4315–4330
28. Aberman K, Eldar YC (2017) Sub-Nyquist SAR via fourier domain range-doppler processing. IEEE Trans Geosci Remote Sens 55(11):6228–6244
29. Zheng L, Wang X (2017) Super-resolution delay-doppler estimation for OFDM passive radar. IEEE Trans Signal Process 65(9):2197–2210

30. Donoho DL, Maleki A, Montanari A (2009) Message passing algorithms for compressed sensing. Proc Natl Acad Sci USA 106(45):18914–18919
31. Rangan S (2011) Generalized approximate message passing for estimation with random linear mixing. In: IEEE International Symposium on Information Theory Proceedings. IEEE, pp 2168–2172
32. Al-Shoukairi M, Schniter P, Rao BD (2018) A GAMP-based low complexity sparse bayesian learning algorithm. IEEE Trans Signal Process 66(2):294–308
33. Vila J, Schniter P (2011) Expectation-maximization Bernoulli-Gaussian approximate message passing. In: Conference record of the Asilomar conference on signals, systems, and computers, pp 799–803
34. Guo C, Nelson JDB (2015) Direction of arrival estimation via approximate message passing. In: IET Conference, pp 1–6
35. Wei C, Liu H, Zhang Z et al (2017) Approximate message passing-based joint user activity and data detection for NOMA. IEEE Commun Lett 21(3):640–643
36. Claudio EDD, Parisi R, Jacovitti G (2018) Space time MUSIC: consistent signal subspace estimation for wideband sensor arrays. IEEE Trans Signal Process 66(10):2685–2699
37. Vallet P, Loubaton P (2017) On the performance of MUSIC with Toeplitz rectification in the context of large arrays. IEEE Trans Signal Process 65(22):5848–5859
38. Liu A, Zhang X, Yang Q et al (2019) Combined root-MUSIC algorithms for multi-carrier MIMO radar with sparse uniform linear arrays. IET Radar Sonar Navig 13(1):89–97

Chapter 6
Two Timescale Robust Energy-Efficient Precoding for Dual-Polarized MIMO Systems

Yongjia Jin, Xue Yin, and Ziyi Yang

In this chapter, we propose the two timescale based dual-structured linear precoding scheme for the dual-polarized multiuser MIMO downlink system. Section 6.1 introduces the research status of EE optimization in a dual-polarized antenna multi-user MIMO downlink communication system. Section 6.2 introduces the system model of the dual-polarized multiuser MIMO downlink. After that, Sect. 6.3 is devoted to our proposed dual-structured linear precoding scheme and polarization-based subgrouping technique for realizing the robust system EE optimization of the dual-polarized multiuser MIMO downlink. Specifically, Sect. 6.3.1 of this section introduces the optimization of the preprocessing matrix based on the dual-polarized array spatial correlation, while the proposed iterative algorithm to optimize the linear precoding is detailed in Sect. 6.3.2. Section 6.4 gives the convergence and complexity analysis of Algorithm 6.1. Furthermore, the simulation results are given in Sect. 6.5, followed by Sect. 6.6 which concludes this chapter.

6.1 Introduction

The spatial physical size-limited nature of MIMO systems tends to cause correlations between adjacent antenna elements, thereby compromising the potential performance gains of MIMO systems [1, 2]. One possible solution to the above issue is to employ

Y. Jin · X. Yin · Z. Yang (✉)
Beijing Institute of Technology, Beijing, China
e-mail: yziyi@bit.edu.cn

Y. Jin
e-mail: jinyongjia@bit.edu.cn

X. Yin
e-mail: yx9039@126.com

X. Ding et al. (eds.), *Performance Analysis and Improvement in MIMO Communication Systems*, https://doi.org/10.1007/978-981-96-5951-7_6

a dual-polarized antenna array [3], because of its cost and space efficiency, in which multiple antennas are grouped in pairs with orthogonal polarizations. Considering that electromagnetic waves experience independent fading in both LoS and non-line-of-sight (NLoS) scenarios, it has been revealed that a low spatial correlation can be achieved by employing the dual-polarized antenna array [2, 4]. Furthermore, in the dual-polarized antenna arrays, both the mutual coupling effect between the antenna elements and the feedback amount of the CSI can be substantially reduced [5].

In view of the above advantages of dual-polarized antenna arrays, this technology has received extensive attention [6–12]. For instance, in [7], a dual-polarized channel was modeled based on the vector geometrical scattering mechanism. The authors in [8] presented an analytical channel model of dual-polarized two-input two-output and MIMO transmissions, which could efficiently separate the spatial and polarization effects. The performance of spatial multiplexing and space-time block coding (e.g., Alamouti coding) in MIMO dual-polarized antenna systems was investigated in [9]. To consider a more practical Ricean fading channel, an adaptive polarization-based scheme was proposed to maximize the desired signal power and minimize the interference signal power in [10]. The authors in [11] designed the codebook that could minimize the average SNR distortion, followed by proposing a limited feedback beamforming scheme for achieving the reduction of the CSI feedback overhead. To further reduce both the implementation complexity and the feedback overhead, the polarized subgrouping technique and the dual-structured linear precoding scheme were utilized in the multi-polarized MIMO system [12]. The above studies have effectively improved the SE performance of the MIMO systems.

In addition to spectral efficiency, EE is also considered an important indicator of the performance of MIMO systems. Specifically, EE has been widely accepted as a critical performance metric for both beamforming and precoding optimizations. Up to now, a rich body of literature has conducted in-depth research on the above technologies. For example, in [13], a power allocation combining scheme with the beamforming technique was employed to maximize the EE performance of the MIMO network by utilizing the uplink-downlink duality property. Meanwhile, an energy-efficient precoding scheme was proposed in [14] for achieving the maximization of weighted sum EEs in the heterogeneous multi-cell multiuser networks. Unlike most existing studies, which have exclusively focused on maximizing the system-wide EE, the authors in [15] investigated the fairness-aware EE optimization by maximizing the minimum attainable EE among all mobile users (MUs) in the multi-cell multiuser networks. More specifically, two iterative methods (i.e., the block coordinate descent method and the sequential convex approximation method) were employed in [15] for solving the problem of maximizing the fairness-aware EE.

Despite this, there are still some shortcomings in the existing works, especially when considering CSI feedback, existing research generally assumes that the perfect CSI is available at the communication devices. However, this assumption is hard to realize, due to the fact that many practical constraints (e.g., feedback errors) could be imposed on the communication systems. Generally speaking, the receiver should estimate the CSI and then feed it back to the transmitter, so both errors negative effects of these errors and uncertainties on system performance, robust design approaches have been proposed, such as robust adaptive beamforming based on interference-plus-noise covariance matrix reconstruction [16, 17], and cross-correlation [18], minimum variance distortionless response beamforming [19], and Bayesian approach [20]. Consequently, there is abundant literature addressing the improvement of the robust EE performance in wireless communication systems [21–24]. Specifically, from the statistical robustness perspective, the authors in [21] designed the optimal EE precoding scheme by relying on the statistical CSI in the two-hop relaying channel. Furthermore, in [22], the trade-off between EE and SE was studied under the statistical CSI case and with the asymmetric traffic requirements in a two-way relaying network. Meanwhile, from the worst-case robustness point of view, the robust EE maximization in underlay cognitive radio networks was investigated based on the fractional programming in [23]. In addition, in [24], the authors investigated the worst-case EE with the multiplicative or additive bounded CSI errors, in which the channel-diagonalizing structure was demonstrated to be optimal for the robust EE precoding design. In recent years, a rate-splitting (RS) approach was proposed, which can outperform conventional communication strategies on EE and SE performance [25–27]. RS could benefit from partial channel state information at the transmitter (CSIT) and tackle the multi-user interference by splitting the messages into common and private parts, then the common stream encoded from the common parts was superposed on top of all private streams encoded from the private parts.

In addition to the above studies, there is some literature on the EE performance of the polarized MIMO systems. For example, the authors in [28] analyzed the EE performance of variant precoded-polarized-MIMO-based architectures using a space-polarization MIMO test-bed, where the accurate and repeatable physical CSI is assumed to be available. In [29], the transmitter EE performance of the dual-polarized MIMO architecture was studied for the packet-based communication systems, in which a closed-form approximation for the EE metric was obtained. Furthermore, the trade-offs between variant performance metrics, including EE, SE, modulation scheme, and packet design, were also investigated [29].

Notation: Throughout this chapter, the following notations are used throughout the chapter. The normal-faced lower-case letters are used to denote scalars. The bold-faced lower-case and upper-case letters are used to denote vectors and matrices, respectively. For matrices, the conjugate, transpose, and conjugate transpose operators are denoted by $(\cdot)^*$, $(\cdot)^{\mathrm{T}}$, and $(\cdot)^{\mathrm{H}}$, respectively. In addition, the symbols $|\cdot|$, $\|\cdot\|$, and $\|\cdot\|_F$ denote the absolute value, Euclidean norm and matrix Frobenius norm, respectively. The mathematical operator $\otimes$ denotes the Kronecker product and $\odot$ denotes the Hadamard product, and moreover $\mathrm{Tr}(\cdot)$ denotes the matrix trace operator. The notation $\boldsymbol{A} \succ \mathbf{0}$ and $\boldsymbol{A} \succeq \mathbf{0}$ indicate that $\boldsymbol{A}$ is positive and semi-positive definite matrix, respectively, where $\mathbf{0}$ is the zero matrix of an appropriate dimension. Similarly, $\boldsymbol{A} \succeq \boldsymbol{B}$ means that $(\boldsymbol{A} - \boldsymbol{B})$ is positive semi-definite. Additionally, the symbol $\mathbb{E}\{\cdot\}$ denotes the expectation operator, and $\boldsymbol{I}_n$ is the $n \times n$ identity matrix, while $\mathbf{1}_{n\times m}$ denotes the $n \times m$ matrix whose elements are all equal to 1.

6.2 System Structure and Signal Model

In this section, we investigate the EE optimization for multiuser MIMO (MU-MIMO) downlink communication systems. In the considered MU-MIMO systems, a BS serves M single horizontally or vertically polarized antenna MUs, and the BS is equipped with N_t polarized antennas. Without loss of generality, in our work, N_t is even and hence the antennas of the BS can be grouped in $N_t/2$ pairs of co-located horizontally and vertically polarized, i.e., dual-polarized antennas. The M MUs are clustered into K groups each composed of the same number $N_u = \frac{M}{K}$ of MUs. For simplicity, we further assume that each MU group has $N_u/2$ vertically polarized MUs and $N_u/2$ horizontally polarized MUs. In the considered MU-MIMO downlink communications, the received signal $\boldsymbol{y}_k \in \mathbb{C}^{N_u}$ of the kth MU group equals

$$\boldsymbol{y}_k = \boldsymbol{H}_k^{\mathrm{H}} \sum_{j \in \mathcal{K}} \boldsymbol{V}_j \boldsymbol{s}_j + \boldsymbol{n}_k, \tag{6.1}$$

where $\mathcal{K} = \{1, 2, \ldots, K\}$ denotes the set of the K MU groups, $\boldsymbol{s}_j \in \mathbb{C}^{N_u}$ is the transmitted symbol vector to the jth MU group with $\mathbb{E}\left\{\boldsymbol{s}_j \boldsymbol{s}_j^{\mathrm{H}}\right\} = \boldsymbol{I}_{N_u}$, $\boldsymbol{V}_j \in \mathbb{C}^{N_t \times N_u}$ is the precoding matrix for the jth MU group, and $\boldsymbol{H}_k \in \mathbb{C}^{N_t \times N_u}$ represents the channel matrix linking the N_t BS antennas to the N_u MUs of the kth group, while $\boldsymbol{n}_k \in \mathbb{C}^{N_u}$ is the corresponding AWGN vector, whose elements obey the Gaussian distribution with zero mean and power σ_n^2 and hence $\mathbb{E}\left\{\boldsymbol{n}_k \boldsymbol{n}_k^{\mathrm{H}}\right\} = \sigma_n^2 \boldsymbol{I}_{N_u}$. Additionally, we assume that $N_t > N_u$.

To explicitly reflect the dual-polarized structures of all the signals involved, $\boldsymbol{y}_k$ is decomposed into vertically and horizontally polarized signals, i.e.,

$$\boldsymbol{y}_k = \left[\left(\boldsymbol{y}_k^{v}\right)^{\mathrm{T}} \left(\boldsymbol{y}_k^{h}\right)^{\mathrm{T}}\right]^{\mathrm{T}}, \tag{6.2}$$

where $\boldsymbol{y}_k^{v} \in \mathbb{C}^{N_u/2}$ and $\boldsymbol{y}_k^{h} \in \mathbb{C}^{N_u/2}$ correspond to the received vertically and horizontally polarized signals of the kth MU group, respectively. Similarly, we have

$$\boldsymbol{s}_j = \left[\left(\boldsymbol{s}_j^{v}\right)^{\mathrm{T}} \left(\boldsymbol{s}_j^{h}\right)^{\mathrm{T}}\right]^{\mathrm{T}}, \tag{6.3}$$

with the vertically and horizontally polarized components $\boldsymbol{s}_j^{v} \in \mathbb{C}^{N_u/2}$ and $\boldsymbol{s}_j^{h} \in \mathbb{C}^{N_u/2}$. Correspondingly, the precoding matrix can be expressed as $\boldsymbol{V}_j = \left[\boldsymbol{V}_j^{v}\ \boldsymbol{V}_j^{h}\right]$, with $\boldsymbol{V}_j^{p} \in \mathbb{C}^{N_t \times \frac{N_u}{2}}$ for $p \in \{v, h\}$. In addition, $\boldsymbol{n}_k = \left[\left(\boldsymbol{n}_k^{v}\right)^{\mathrm{T}} \left(\boldsymbol{n}_k^{h}\right)^{\mathrm{T}}\right]^{\mathrm{T}}$ with $\boldsymbol{n}_j^{v} \in \mathbb{C}^{N_u/2}$ and $\boldsymbol{n}_j^{h} \in \mathbb{C}^{N_u/2}$.

As a result, the signal model (6.1) can be rewritten as

$$\begin{bmatrix} \boldsymbol{y}_k^{v} \\ \boldsymbol{y}_k^{h} \end{bmatrix} = \begin{bmatrix} \left(\boldsymbol{H}_k^{vv}\right)^{\mathrm{H}} & \left(\boldsymbol{H}_k^{vh}\right)^{\mathrm{H}} \\ \left(\boldsymbol{H}_k^{hv}\right)^{\mathrm{H}} & \left(\boldsymbol{H}_k^{hh}\right)^{\mathrm{H}} \end{bmatrix} \sum_{j \in \mathcal{K}} \boldsymbol{V}_j \begin{bmatrix} \boldsymbol{s}_j^{v} \\ \boldsymbol{s}_j^{h} \end{bmatrix} + \begin{bmatrix} \boldsymbol{n}_k^{v} \\ \boldsymbol{n}_k^{h} \end{bmatrix}, \tag{6.4}$$

where $\boldsymbol{H}_k^{mn} \in \mathbb{C}^{\frac{N_t}{2} \times \frac{N_u}{2}}$, $\forall m, n \in \{h, v\}$, denotes the channel matrix between the m-polarized MUs in the kth group and the n-polarized BS antennas. For the polarized channel model, the spatial covariance matrices for the horizontally and vertically polarized channels are usually almost the same [12], we can use the same polarized spatial covariance matrix $\boldsymbol{R}_k \in \mathbb{C}^{\frac{N_t}{2} \times \frac{N_t}{2}}$ for representing these two spatial covariance matrices of the kth group of MUs. Moreover, the EVD of $\boldsymbol{R}_k$ is defined in the following

$$\boldsymbol{R}_k = \boldsymbol{U}_k \boldsymbol{\Lambda}_k \boldsymbol{U}_k^{\mathrm{H}}, \tag{6.5}$$

where the diagonal matrix $\boldsymbol{\Lambda}_k$ contains the r_k non-zero eigenvalues of $\boldsymbol{R}_k$ arranged in the descending order, and $\boldsymbol{U}_k \in \mathbb{C}^{\frac{N_t}{2} \times r_k}$ is composed of the r_k eigenvectors corresponding to the eigenvalues in $\boldsymbol{\Lambda}_k$. Generally, we have $r_k < \frac{N_t}{2}$. According to [12], the dual-polarized MIMO channel $\boldsymbol{H}_k$ of the kth group can be written as

$$\boldsymbol{H}_k = \begin{bmatrix} \boldsymbol{H}_k^{vv} & \boldsymbol{H}_k^{hv} \\ \boldsymbol{H}_k^{vh} & \boldsymbol{H}_k^{hh} \end{bmatrix} = \left(\begin{bmatrix} 1 & \gamma_p \\ \gamma_p & 1 \end{bmatrix} \otimes \left(\boldsymbol{U}_k \boldsymbol{\Lambda}_k^{\frac{1}{2}} \right) \right) \times \left(\boldsymbol{G}_k \odot \left(\boldsymbol{X} \otimes \mathbf{1}_{r_k \times \frac{N_u}{2}} \right) \right), \tag{6.6}$$

where γ_p reflects the correlation between the horizontally and vertically polarized antennas, which is generally close to 0, and $\boldsymbol{G}_k \in \mathbb{C}^{2r_k \times N_u}$ and $\boldsymbol{X} \in \mathbb{R}^{2\times 2}$ are expressed respectively as follow [12]

$$\boldsymbol{G}_k = \begin{bmatrix} \boldsymbol{G}_k^{vv} & \boldsymbol{G}_k^{hv} \\ \boldsymbol{G}_k^{vh} & \boldsymbol{G}_k^{hh} \end{bmatrix}, \quad \boldsymbol{X} = \begin{bmatrix} 1 & \sqrt{\chi} \\ \sqrt{\chi} & 1 \end{bmatrix}, \tag{6.7}$$

where $\boldsymbol{G}_k$ is the short-term channel fading matrix with $\boldsymbol{G}_k^{mn} \in \mathbb{C}^{r_k \times \frac{N_u}{2}}$, $\forall m, n \in \{h, v\}$, whose elements follow independently and identically complex Gaussian distributed with zero mean and variance $\sigma^2 = 1$, and $\boldsymbol{X}$ with $0 \le \chi \le 1$ represents the channel ability to distinguish horizontal and vertical polarizations [30]. In this section, considering the ideal case that the orthogonally polarized signals can be separated from each other perfectly, i.e., $\chi \approx 0$. Note that the long-term parameter $\boldsymbol{R}_k$ is slowly varying and assumed to be obtained accurately with a low feedback overhead. However, the short-term CSI $\boldsymbol{G}_k$ changes rapidly, its feedback to the BS is generally imperfect due to many factors, such as feedback errors and quantization. Accordingly, when channel errors are taken into account, the channel matrix $\boldsymbol{G}_k$ satisfies the following model

$$\boldsymbol{G}_k = \widehat{\boldsymbol{G}}_k + \boldsymbol{\Delta}_k, \ \|\boldsymbol{\Delta}_k\|_F \le \sigma_k, \forall k \in \mathcal{K}, \tag{6.8}$$

where $\widehat{\boldsymbol{G}}_k$ denotes the estimated nominal channel, and $\boldsymbol{\Delta}_k$ is the CSI error which is norm bounded by the threshold σ_k.

In order to formulate the EE criterion of the dual-polarized MIMO system, based on the signal model (6.4), the received signal $y_{k,i}^p$ at the ith MU in the kth group with p-polarization can be expressed as

$$
\begin{aligned}
y_{k,i}^{p} = \boldsymbol{H}_{k,i}^{p}\boldsymbol{V}_{k}^{p,i}s_{k,i}^{p} + & \sum_{l\in\left\{1,\ldots,\frac{N_u}{2}\right\},\, l\neq i} \boldsymbol{H}_{k,i}^{p}\boldsymbol{V}_{k}^{p,l}s_{k,l}^{p} \\
& + \boldsymbol{H}_{k,i}^{p}\boldsymbol{V}_{k}^{q}\boldsymbol{s}_{k}^{q} + \sum_{j\in\mathcal{K}\backslash k} \boldsymbol{H}_{k,i}^{p}\boldsymbol{V}_{j}\boldsymbol{s}_{j} + n_{k,i}^{p}, \\
\forall i \in \left\{1, 2, \ldots, \frac{N_u}{2}\right\}&,\ p, q \in \{h, v\}, q \neq p,\ k \in \mathcal{K},
\end{aligned}
\tag{6.9}
$$

where $\boldsymbol{H}_{k,i}^{v}$ and $\boldsymbol{H}_{k,i}^{h}$ denote the ith rows of $\left[\left(\boldsymbol{H}_{k}^{vv}\right)^{\mathrm{H}}\ \left(\boldsymbol{H}_{k}^{vh}\right)^{\mathrm{H}}\right]$ and $\left[\left(\boldsymbol{H}_{k}^{hv}\right)^{\mathrm{H}}\ \left(\boldsymbol{H}_{k}^{hh}\right)^{\mathrm{H}}\right]$, respectively. The vector $\boldsymbol{V}_{k}^{p,i}$ is the ith column of the precoding matrix $\boldsymbol{V}_{k}^{p}$. On the other hand, $s_{k,i}^{p}$ and $n_{k,i}^{p}$ are the ith elements of $\boldsymbol{s}_{k}^{p}$ and $\boldsymbol{n}_{k}^{p}$, respectively. The system EE metric for the multiuser MIMO downlink communication can then be defined as the ratio of the system achievable sum rate to the total power consumption of the system [31], which is of the following mathematical formula

$$
\mathrm{EE} = \frac{\sum\limits_{i\in\left\{1,\ldots,\frac{N_u}{2}\right\},k\in\mathcal{K},p\in\{h,v\}} R_{k,i}^{p}}{\sum\limits_{i\in\left\{1,\ldots,\frac{N_u}{2}\right\},k\in\mathcal{K},p\in\{h,v\}} \left\|\boldsymbol{V}_{k}^{p,i}\right\|^{2} + P_c}. \tag{6.10}
$$

In (6.10), the data rate of the ith MU in the kth group with p-polarization is

$$
R_{k,i}^{p} = \log\left(1 + \frac{\left|\boldsymbol{H}_{k,i}^{p}\boldsymbol{V}_{k}^{p,i}\right|^{2}}{\sigma_n^2 + \sum\limits_{l}\left|\boldsymbol{H}_{k,i}^{p}\boldsymbol{V}_{k}^{p,l}\right|^{2} + \left\|\boldsymbol{H}_{k,i}^{p}\boldsymbol{V}_{k}^{q}\right\|^{2} + \sum\limits_{j}\left\|\boldsymbol{H}_{k,i}^{p}\boldsymbol{V}_{j}\right\|^{2}}\right), \tag{6.11}
$$

where $l \in \left\{1, \ldots, \frac{N_u}{2}\right\}$, $l \neq i$, $j \in \mathcal{K}\backslash k$, $q \in \{h, v\}$, $q \neq p$, and P_c denotes the circuit power consumption of the BS, which is considered as a constant in this chapter [32].

When the norm-bounded channel error model in (6.8) is taken into account, the proposed robust MU-MIMO precoding optimization aims at maximizing the worst-case achievable system EE. Based on (6.10), the robust system EE optimization problem is finally formulated as follows

$$
\begin{aligned}
&\max_{\boldsymbol{V}_k,k\in\mathcal{K}}\ \min_{\boldsymbol{\Delta}_k,k\in\mathcal{K}}\ \mathrm{EE}, \\
&\text{s.t.}\ \sum_{k\in\mathcal{K}} \mathrm{Tr}\left(\boldsymbol{V}_k\boldsymbol{V}_k^{\mathrm{H}}\right) \leq P_{\max},
\end{aligned}
\tag{6.12}
$$

where $P_{\max}$ is the maximum transmit power. The problem (6.12) aims at maximizing the worst-case system EE with the imperfect CSI. However, due to the nonconvexity of the optimization problem (6.12), it is very challenging to directly solve this non-convex fractional optimization problem.

6.3 Two Timescale Based Dual-Structured Linear Precoding

In this chapter, the dual-structured linear precoding is adopted [33]. Then the precoding matrix $\boldsymbol{V}_k$ for the MUs of the kth group can be written as

$$\boldsymbol{V}_k = \boldsymbol{B}_k \boldsymbol{P}_k, \ \forall k \in \mathcal{K}, \tag{6.13}$$

where $\boldsymbol{B}_k \in \mathbb{C}^{N_t \times L}$ is the preprocessing matrix based on the long-term CSI, and $\boldsymbol{P}_k \in \mathbb{C}^{L \times N_u}$ is the instantaneous CSI based linear precoding matrix, i.e., the precoding design is based on two timescale CSI $\boldsymbol{R}_k$ and $\boldsymbol{G}_k$, while L, satisfying $N_u \leq L \leq 2r_k < N_t$, defines the dimension of the transformed equivalent channel $\boldsymbol{H}_k^{\mathrm{H}} \boldsymbol{B}_k$. Since the long-term CSI can be accurately acquired, $\boldsymbol{B}_k$ can be optimized utilizing the accurate polarized spatial correlation matrix $\boldsymbol{R}_k$, while the optimization of $\boldsymbol{P}_k$ needs to consider the CSI error. Although the adopted dual-structured precoding scheme $\boldsymbol{V}_k = \boldsymbol{B}_k \boldsymbol{P}_k$ will cause the suboptimality of EE, this design significantly reduces both CSI feedback overhead and EE optimization complexity.

6.3.1 *The Long-Term CSI Based Preprocessing Matrix*

To cancel the interference to the MUs in other groups as much as possible, the preprocessing matrix $\boldsymbol{B}_k$ for the kth group should satisfy $\boldsymbol{H}_j^{\mathrm{H}} \boldsymbol{B}_k \approx \boldsymbol{0}, \forall j, k \in \mathcal{K}, \ j \neq k$. To realize the objective above, we consider the polarized spatial covariance matrix $\boldsymbol{R}_k$ of (6.5). We firstly define the signal space matrix $\boldsymbol{U}_{-k} \in \mathbb{C}^{\frac{N_t}{2} \times \sum_{j \in \mathcal{K} \setminus k} r_j^a}$ for the kth MU group as

$$\boldsymbol{U}_{-k} = \left[\widetilde{\boldsymbol{U}}_1 \cdots \widetilde{\boldsymbol{U}}_{k-1} \ \widetilde{\boldsymbol{U}}_{k+1} \cdots \widetilde{\boldsymbol{U}}_K \right], \tag{6.14}$$

where $\widetilde{\boldsymbol{U}}_k$ contains the first r_k^a columns of $\boldsymbol{U}_k$ with $r_k^a \leq r_k$. The parameter r_k^a determines the number of the dominant eigenvalues of $\boldsymbol{R}_k$. Moreover, the SVD of $\boldsymbol{U}_{-k}$ is defined in the following

$$\boldsymbol{U}_{-k} = \left[\boldsymbol{Q}_{-k}^0 \ \boldsymbol{Q}_{-k}^1 \right] \begin{bmatrix} \boldsymbol{\Lambda}_{-k}^0 \\ \boldsymbol{0} \end{bmatrix} \boldsymbol{V}_{-k}^{\mathrm{H}}, \tag{6.15}$$

where the diagonal matrix $\boldsymbol{\Lambda}_{-k}^0$ contains of $\sum_{j \in \mathcal{K} \setminus k} r_j^a$ dominant singular values of $\boldsymbol{U}_{-k}$, the matrices $\boldsymbol{Q}_{-k}^0$ and $\boldsymbol{Q}_{-k}^1$ consist of the left singular vectors of $\boldsymbol{U}_{-k}$, and $\boldsymbol{V}_{-k}$ contains the right singular vectors of $\boldsymbol{U}_{-k}$.

Based on the SVD in (6.15), the orthogonal space of $\boldsymbol{U}_{-k}$ can be found as $\boldsymbol{Q}_{-k}^{1} \in \mathbb{C}^{\frac{N_t}{2} \times \left(\frac{N_t}{2} - \sum_{j \in \mathcal{K}\backslash k} r_j^a\right)}$, which satisfies $\left(\boldsymbol{Q}_{-k}^{1}\right)^{\mathrm{H}} \boldsymbol{U}_{-k} = \boldsymbol{0}$. In addition, based on (6.6) and the fact that $\chi \approx 0$ and $\gamma_p \approx 0$, the spatial covariance matrix of $\boldsymbol{H}_k$ is given by

$$\widetilde{\boldsymbol{R}}_k = \mathrm{Blockdiag}\{\boldsymbol{R}_k, \boldsymbol{R}_k\}. \tag{6.16}$$

Due to the block diagonal structure of $\widetilde{\boldsymbol{R}}_k$, the optimal long-term CSI based $\boldsymbol{B}_k$ is also block diagonal. Specifically, we have

$$\boldsymbol{B}_k = \boldsymbol{I}_2 \otimes \boldsymbol{B}_k^s, \quad \boldsymbol{B}_k^s = \boldsymbol{Q}_{-k}^{1} \widetilde{\boldsymbol{F}}_k, \tag{6.17}$$

where $\widetilde{\boldsymbol{F}}_k \in \mathbb{C}^{\left(\frac{N_t}{2} - \sum_{j \in \mathcal{K}\backslash k} r_j^a\right) \times \frac{L}{2}}$ is derived from the effective channel $\widetilde{\boldsymbol{H}}_k = \left(\boldsymbol{I}_2 \otimes \boldsymbol{Q}_{-k}^{1}\right)^{\mathrm{H}} \boldsymbol{H}_k$ of the kth MU group as follows. First, the covariance matrix of $\widetilde{\boldsymbol{H}}_k$ is given by

$$\overline{\boldsymbol{R}}_k = \mathrm{diag}\{\left(\boldsymbol{Q}_{-k}^{1}\right)^{\mathrm{H}} \boldsymbol{R}_k \boldsymbol{Q}_{-k}^{1}, \left(\boldsymbol{Q}_{-k}^{1}\right)^{\mathrm{H}} \boldsymbol{R}_k \boldsymbol{Q}_{-k}^{1}\}. \tag{6.18}$$

Next, we obtain the following EVD

$$\left(\boldsymbol{Q}_{-k}^{1}\right)^{\mathrm{H}} \boldsymbol{R}_k \boldsymbol{Q}_{-k}^{1} = \boldsymbol{F}_k \widetilde{\boldsymbol{\Lambda}}_k \boldsymbol{F}_k^{\mathrm{H}}. \tag{6.19}$$

Obviously, the rank of $\left(\boldsymbol{Q}_{-k}^{1}\right)^{\mathrm{H}} \boldsymbol{R}_k \boldsymbol{Q}_{-k}^{1}$ satisfies

$$\mathrm{rank}\left(\left(\boldsymbol{Q}_{-k}^{1}\right)^{\mathrm{H}} \boldsymbol{R}_k \boldsymbol{Q}_{-k}^{1}\right) = \min\left\{\frac{N_t}{2} - \sum_{j \in \mathcal{K}\backslash k} r_j^a, r_k\right\}. \tag{6.20}$$

Hence, the unitary matrix $\boldsymbol{F}_k$ consists of the $\min\left\{\frac{N_t}{2} - \sum_{j \in \mathcal{K}\backslash k} r_j^a, r_k\right\}$ dominant eigenvectors associated with $\left(\boldsymbol{Q}_{-k}^{1}\right)^{\mathrm{H}} \boldsymbol{R}_k \boldsymbol{Q}_{-k}^{1}$. Therefore, we can use the first $\frac{L}{2}$ columns of $\boldsymbol{F}_k$ to form $\widetilde{\boldsymbol{F}}_k$, assuming that $\frac{L}{2} \leq \min\left\{\frac{N_t}{2} - \sum_{j \in \mathcal{K}\backslash k} r_j^a, r_k\right\}$. Using this $\widetilde{\boldsymbol{F}}_k$ in (6.17), we obtain the optimal preprocessing matrix $\boldsymbol{B}_k$ for the MUs in the kth group.

It can be found that the transmit signals for the kth group of MUs are projected onto the L-dimensional dominant eigenspace, which is orthogonal to the partial channel eigenspaces of the other non-intended groups, i.e., $\boldsymbol{H}_k^{\mathrm{H}} \boldsymbol{B}_j \approx \boldsymbol{0}, \forall j, k \in \mathcal{K}$ and $j \neq k$. We note that only when the value of r_k^a approximates r_k, the perfect orthogonality can be guaranteed. Because

$$\boldsymbol{F}_k \in \mathbb{C}^{\left(\frac{N_t}{2} - \sum_{j \in \mathcal{K}\backslash k} r_j^a\right) \times \min\left\{\frac{N_t}{2} - \sum_{j \in \mathcal{K}\backslash k} r_j^a, r_k\right\}}, \tag{6.21}$$

and $\widetilde{\boldsymbol{F}}_k$ contains the first $\frac{L}{2}$ columns of $\boldsymbol{F}_k$, the values of r_k^a and L must be selected to satisfy the following constraint

$$N_u \leq L \leq \min\left\{N_t - 2\sum_{j\in\mathcal{K}\setminus k} r_j^a, 2r_k\right\}, \forall k \in \mathcal{K}. \tag{6.22}$$

Without loss of generally, in the following for r_k^a in (6.22), it is assumed that $r_1^a = \cdots = r_K^a = r$. As stated above, if r increases to $\min\limits_{\forall k\in\mathcal{K}} r_k$, the better orthogonality of the preprocessing matrix $\boldsymbol{B}_k$ (i.e., $\boldsymbol{H}_j^{\mathrm{H}}\boldsymbol{B}_k = \boldsymbol{0}$) is achieved, and thus more interference to the MUs in the non-intended groups can be mitigated, especially when $r_1 = \cdots = r_K$ holds, this inter-group interference can be perfectly canceled. Unfortunately, the range of L becomes smaller when r increases according to (6.22), which unavoidably reduces the design freedom of the linear precoder $\boldsymbol{P}_k$. It can be seen that the value of r reveals a trade-off between the perfect orthogonality of the preprocessing matrix $\boldsymbol{B}_k$ and the degree of freedom in the optimization for the linear precoding $\boldsymbol{P}_k$. This influence will be further investigated in the simulation study.

Furthermore, the preprocessing matrix $\boldsymbol{B}_k$ obtained above using the block diagonalization also realizes the maximum achievable rate of the corresponding users given the linear precoder $\boldsymbol{P}_k$ [33]. Specifically, it is clear that based on (6.17) as well as the formulations of $\boldsymbol{Q}_{-k}^1$ and $\widetilde{\boldsymbol{F}}_k$, we have

$$\begin{aligned}\boldsymbol{B}_k^{\mathrm{H}}\boldsymbol{B}_k &= \boldsymbol{I}_2 \otimes \widetilde{\boldsymbol{F}}_k^{\mathrm{H}}\left(\boldsymbol{Q}_{-k}^1\right)^{\mathrm{H}}\boldsymbol{Q}_{-k}^1\widetilde{\boldsymbol{F}}_k \\ &= \boldsymbol{I}_2 \otimes \boldsymbol{I}_{\frac{L}{2}} = \boldsymbol{I}_L.\end{aligned} \tag{6.23}$$

By expressing $\boldsymbol{P}_k = \left[\widetilde{\boldsymbol{P}}_k^v\ \widetilde{\boldsymbol{P}}_k^h\right]$, $\boldsymbol{V}_k^{p,i}$ in (6.9) can be expressed as $\boldsymbol{V}_k^{p,i} = \boldsymbol{B}_k\widetilde{\boldsymbol{P}}_k^{p,i}$, where $\widetilde{\boldsymbol{P}}_k^{p,i}$ is the ith column of $\widetilde{\boldsymbol{P}}_k^p$ for $p \in \{h, v\}$. Therefore the system EE (6.10) can be derived to be

$$\begin{aligned}\mathrm{EE} &= \frac{\sum\limits_{i\in\{1,\dots,\frac{N_u}{2}\},k\in\mathcal{K},p\in\{h,v\}} R_{k,i}^p}{\sum\limits_{i\in\{1,\dots,\frac{N_u}{2}\},k\in\mathcal{K},p\in\{h,v\}} \left\|\boldsymbol{B}_k\boldsymbol{P}_k^{p,i}\right\|^2 + P_c} \\ &= \frac{\sum\limits_{i\in\{1,\dots,\frac{N_u}{2}\},k\in\mathcal{K},p\in\{h,v\}} R_{k,i}^p}{\sum\limits_{i\in\{1,\dots,\frac{N_u}{2}\},k\in\mathcal{K},p\in\{h,v\}} \left\|\widetilde{\boldsymbol{P}}_k^{p,i}\right\|^2 + P_c},\end{aligned} \tag{6.24}$$

where the second equality is due to (6.23). Obviously, the denominator of EE is independent of $\boldsymbol{B}_k$, while its numerator is maximized by the obtained preprocessing matrix $\boldsymbol{B}_k$. Hence, with this optimal $\boldsymbol{B}_k$, the robust EE optimization problem (6.12) is transformed into the optimization problem with a single variable, i.e., the linear precoder $\boldsymbol{P}_k$. As a result, the preprocessing design is beneficial to the EE performance of the system.

6.3.2 The Instantaneous CSI Based Linear Precoding

In most existing works, the linear precoder is configured as a specific form, such as zero-forcing precoder [12]. In this part, we will optimize the linear precoding matrix to achieve the maximization of system EE. As mentioned above, we consider that the values of χ and γ_p are both close to 0. Thus the perfect orthogonality between polarized signals can be obtained, and the interference signals from the cross-polarized channels can be perfectly canceled out. In particular, the channel matrix $\boldsymbol{H}_k$ is block diagonal, i.e.,

$$\boldsymbol{H}_k \approx \text{Blockdiag}\{\boldsymbol{H}_k^{vv}, \boldsymbol{H}_k^{hh}\}. \tag{6.25}$$

Additionally, exploiting the subgrouping technique that separates the MUs into two subgroups with vertical and horizontal polarization, the linear precoding matrix $\boldsymbol{P}_k \in \mathbb{C}^{L\times N_u}$ can be rewritten as

$$\boldsymbol{P}_k = \left[\widetilde{\boldsymbol{P}}_k^v \; \widetilde{\boldsymbol{P}}_k^h\right] = \begin{bmatrix} \boldsymbol{P}_k^v & \boldsymbol{0} \\ \boldsymbol{0} & \boldsymbol{P}_k^h \end{bmatrix}, \tag{6.26}$$

where $\boldsymbol{P}_k^p \in \mathbb{C}^{\frac{L}{2}\times\frac{N_u}{2}}$ for $p \in \{v, h\}$ is related precoding matrix with the p-polarization. From (6.26), we find that the precoding operations are performed for the vertically and horizontally polarized MUs subgroups separately. To be specific, we can design the polarization related $\boldsymbol{P}_k^p$ for $p \in \{h, v\}$ instead of $\boldsymbol{P}_k$, $\forall k \in \mathcal{K}$, which is based on the reduced-dimensional instantaneous CSI. Consequently, the complexity for optimizing $\boldsymbol{P}_k$ is naturally reduced.

Based on the channel matrix $\boldsymbol{H}_k$ and (6.26) as well as the optimal preprocessing matrix $\boldsymbol{B}_k$ derived in Sect. 6.3.1, the received signal of each MU given in (6.9) can be expressed as

$$\begin{aligned} y_{k,i}^p = \; & \boldsymbol{h}_{kk,i}^p \boldsymbol{p}_k^{p,i} s_{k,i}^p + \sum_{l\in\left\{1,\ldots,\frac{N_u}{2}\right\},l\neq i} \boldsymbol{h}_{kk,i}^p \boldsymbol{p}_k^{p,l} s_{k,l}^p \\ & + \sum_{j\in\mathcal{K}\backslash k} \boldsymbol{h}_{kj,i}^p \boldsymbol{P}_j^p \boldsymbol{s}_j^p + n_{k,i}^p, \\ & \forall i \in \left\{1, 2, \ldots, \frac{N_u}{2}\right\},\; p \in \{h, v\}, k \in \mathcal{K}, \end{aligned} \tag{6.27}$$

where $\boldsymbol{p}_k^{p,i}$ is the ith column of the precoding matrix $\boldsymbol{P}_k^p$, for $p \in \{h, v\}$, and $\boldsymbol{h}_{kj,i}^p$ is the ith row of the virtual channel $\left(\boldsymbol{H}_k^{pp}\right)^{\mathrm{H}} \boldsymbol{B}_j^s$. Then, the data rate at each MU can be rewritten as

$$R_{k,i}^p = \log\left(1 + \mathrm{SINR}_{k,i}^p\right), \tag{6.28}$$

and the signal-to-interference plus noise ratio (SINR) is expressed as

$$\mathrm{SINR}_{k,i}^p = \frac{\left|\boldsymbol{h}_{kk,i}^p \boldsymbol{p}_k^{p,i}\right|^2}{\sum\limits_{l} \left|\boldsymbol{h}_{kk,i}^p \boldsymbol{p}_k^{p,l}\right|^2 + \sum\limits_{j \in \mathcal{K} \backslash k} \left\|\boldsymbol{h}_{kj,i}^p \boldsymbol{P}_j^p\right\|^2 + \sigma_n^2}, \tag{6.29}$$

where $l \in \left\{1, \ldots, \frac{N_u}{2}\right\}, l \neq i$. The corresponding user power consumption can be derived to be

$$\mathrm{PW}_{k,i}^p = \mathrm{Tr}\left(\boldsymbol{B}_k^s \boldsymbol{p}_k^{p,i} \left(\boldsymbol{B}_k^s \boldsymbol{p}_k^{p,i}\right)^{\mathrm{H}}\right) = \left\|\boldsymbol{p}_k^{p,i}\right\|^2. \tag{6.30}$$

Thus the resulting system EE of the considered polarized MIMO system can be derived to be

$$\mathrm{EE} = \frac{\sum\limits_{i \in \left\{1, \ldots, \frac{N_u}{2}\right\}, k \in \mathcal{K}, p \in \{h, v\}} R_{k,i}^p}{\sum\limits_{i \in \left\{1, \ldots, \frac{N_u}{2}\right\}, k \in \mathcal{K}, p \in \{h, v\}} \mathrm{PW}_{k,i}^p + P_C}. \tag{6.31}$$

Following that, the norm bounded channel CSI error model in (6.8) is taken into account for the virtual channel $\left(\boldsymbol{H}_k^{pp}\right)^{\mathrm{H}} \boldsymbol{B}_j^s$ and then we have

$$\boldsymbol{h}_{kj,i}^p = \widehat{\boldsymbol{h}}_{kj,i}^p + \boldsymbol{\Delta}_{kj,i}^p, \left\|\boldsymbol{\Delta}_{kj,i}^p\right\| \leq \tau_{kj,i}^p, \ p \in \{h, v\}, i \in \left\{1, 2, \ldots, \frac{N_u}{2}\right\}, \ k, j \in \mathcal{K}, \tag{6.32}$$

where $\widehat{\boldsymbol{h}}_{kj,i}^p$ is the estimated nominal channel, and $\boldsymbol{\Delta}_{kj,i}^p$ is the related CSI error for $\boldsymbol{h}_{kj,i}^p \in \mathbb{C}^{1 \times \frac{L}{2}}$, while the channel error threshold $\tau_{kj,i}^p$ satisfies

$$\tau_{kj,i}^p - \sqrt{\frac{\mathrm{Tr}(\boldsymbol{R}_k)}{2N_u}} \sigma_k \left\|\boldsymbol{B}_j^s\right\|_F. \tag{6.33}$$

Based on (6.31) and (6.32), the optimization problem (6.12) can be reformulated as follows

$$\begin{aligned} &\max_{\boldsymbol{P}_k^p, k \in \mathcal{K}, p \in \{h, v\}} \ \min_{\boldsymbol{\Delta}_{\mathcal{K}}, k \in \mathcal{K}} \ \mathrm{EE}, \\ &\text{s.t.} \sum_{k \in \mathcal{K}} \sum_{p \in \{h, v\}} \left\|\boldsymbol{P}_k^p\right\|_F^2 \leq P_{\max}, \end{aligned} \tag{6.34}$$

where the power constraints equals

$$\left\|\boldsymbol{P}_k^p\right\|_F^2 = \mathrm{Tr}\left(\boldsymbol{P}_k^p \left(\boldsymbol{P}_k^p\right)^{\mathrm{H}}\right) = \mathrm{Tr}\left(\boldsymbol{B}_k^s \boldsymbol{P}_k^p \left(\boldsymbol{P}_k^p\right)^{\mathrm{H}} \left(\boldsymbol{B}_k^s\right)^{\mathrm{H}}\right). \tag{6.35}$$

In addition, $\Delta_{\mathcal{K}}$ in the objective function is defined as

$$\boldsymbol{\Delta}_{\mathcal{K}} = \left\{ \boldsymbol{\Delta}_{kj,i}^{p} : \forall i \in \left\{1, \ldots, \frac{N_u}{2}\right\}, k, j \in \mathcal{K}, p \in \{h, v\} \right\}, \tag{6.36}$$

which contains all the CSI uncertainties. It is worth noting that based on the following equality

$$\min_{\boldsymbol{\Delta}_{\mathcal{K}}, k \in \mathcal{K}} \ \mathrm{EE} = \frac{\min_{\boldsymbol{\Delta}_{\mathcal{K}}} \sum_{k,p,i} R_{k,i}^{p}}{\sum_{k,p,i} \mathrm{PW}_{k,i}^{p} + P_C}, \tag{6.37}$$

thus the optimization problem (6.34) can be transformed to the equivalent problem as follows [34]

$$\begin{aligned} &\max_{\boldsymbol{P}_k^p, \eta} G(\eta), \\ &\text{s.t. } \|\boldsymbol{P}\|_F^2 \leq P_{\max}, \end{aligned} \tag{6.38}$$

where $\boldsymbol{P} = \left[\boldsymbol{P}_1^h \ \boldsymbol{P}_1^v \cdots \boldsymbol{P}_K^h \ \boldsymbol{P}_K^v\right] \in \mathbb{C}^{\frac{L}{2} \times K N_u}$ and the objective function $G(\eta)$ is defined as

$$G(\eta) = \min_{\boldsymbol{\Delta}_{\mathcal{K}}} \sum_{k,p,i} R_{k,i}^{p} - \eta \sum_{k,p,i} \mathrm{PW}_{k,i}^{p} - \eta P_C, \tag{6.39}$$

which is monotonically decreasing with respect to η. Following the similar logic as given in [35], it can be concluded that the optimal precoding matrix for (6.38) is achieved from

$$\max_{\boldsymbol{P}_k^p} \ G(\eta) = 0. \tag{6.40}$$

Furthermore, it can be proved that the range of η is

$$\eta \in \left[0, \ \frac{\sum_{i,k,p} R_{k,i}^{p \max}}{P_C} \right], \tag{6.41}$$

where

$$\begin{aligned} R_{k,i}^{p \max} &= \max_{\|\boldsymbol{p}_k^{p,i}\|^2 \leq P_{\max}} \log\left(1 + \left|\left(\widehat{\boldsymbol{h}}_{kk,i}^{p} + \boldsymbol{\Delta}_{kk,i}^{p}\right)\boldsymbol{p}_k^{p,i}\right|^2 / \sigma_n^2\right) \\ &= \log\left(1 + P_{\max}\left(\left\|\widehat{\boldsymbol{h}}_{kk,i}^{p}\right\| + \tau_{kk,i}^{p}\right)^2 / \sigma_n^2\right). \end{aligned} \tag{6.42}$$

It is noted that $R_{k,i}^{p \max}$ is realized by neglecting the interference and assuming the maximum BS transmit power. Clearly, we can find the optimal η via a one-dimensional search in the range defined by (6.41).

The optimization (6.38) is still a nonconvex problem, because of its nonlinear optimization objective, even given η. In order to overcome the nonconvexity of the considered optimization, the objective function $G(\eta)$ can be rewritten as

$$-G(\eta) = \max_{\mathbf{\Delta}_{\mathcal{K}}} \sum_{i,k,p} \log(\frac{1}{1+\mathrm{SINR}_{k,i}^{p}}) + \eta \sum_{i,k,p} \|\boldsymbol{p}_k^{p,i}\|^2 + \eta P_c, \tag{6.43}$$

We further exploit the relationship between the MSE of the desired signal and the SINR [36, 37] to transform the objective function (6.43). Specifically, let $f_{k,i}^{p}$ be the equalizer of the ith MU in the kth group with p-polarization. Then the MSE is expressed as

$$\begin{aligned}\mathrm{MSE}_{k,i}^{p} &= \mathbb{E}\left\{\left|s_{k,i}^{p} - f_{k,i}^{p} y_{k,i}^{p}\right|^2\right\} \\ &= \|f_{k,i}^{p}\boldsymbol{h}_{kk,i}^{p}\boldsymbol{P}_k^{p} - \boldsymbol{e}_i\|^2 + \sum_{j\in\mathcal{K}\backslash k} \|f_{k,i}^{p}\boldsymbol{h}_{kj,i}^{p}\boldsymbol{P}_j^{p}\|^2 + \sigma_n^2|f_{k,i}^{p}|^2,\end{aligned} \tag{6.44}$$

where $\boldsymbol{e}_i \in \mathbb{R}^{1\times\frac{N_u}{2}}$ is the $\frac{N_u}{2}$-dimensional row vector, and its ith element is 1 while other elements are all equal to zero. It is well-known that the relationship between $\mathrm{SINR}_{k,i}^{p}$ and $\mathrm{MSE}_{k,i}^{p}$ satisfying the following equality [36, 37]

$$\frac{1}{1+\mathrm{SINR}_{k,i}^{p}} = \min_{f_{k,i}^{p}} \mathrm{MSE}_{k,i}^{p}. \tag{6.45}$$

Then the objective function (6.43) can be reformulated as

$$-G(\eta) = \max_{\mathbf{\Delta}_{\mathcal{K}}} \sum_{k,i,p} \log(\min_{f_{k,i}^{p}} \mathrm{MSE}_{k,i}^{p}) + \eta \sum_{k,i,p} \|\boldsymbol{p}_k^{p,i}\|^2 + \eta P_c. \tag{6.46}$$

Substituting (6.46) into (6.38), the optimization problem (6.38) becomes

$$\begin{aligned}\min_{\boldsymbol{P}_k^{p},\eta}\ &\max_{\mathbf{\Delta}_{\mathcal{K}}} \sum_{k,i,p} \log(\min_{f_{k,i}^{p}} \mathrm{MSE}_{k,i}^{p}) + \eta \sum_{k,i,p} \|\boldsymbol{p}_k^{p,i}\|^2 + \eta P_c, \\ \text{s.t.}\ &\|\boldsymbol{P}\|_F^2 \le P_{\max},\end{aligned} \tag{6.47}$$

where $\forall i \in \left\{1, \ldots, \frac{N_u}{2}\right\}, k \in \mathcal{K}, p \in \{h, v\}, \eta$. To further simplify the above problem, we move the operation $\max_{\mathbf{\Delta}_{\mathcal{K}}} \min_{f_{k,i}^{p}}$ outside the log function and utilize the fact that $\max_{y} \min_{x} f(x, y) \le \min_{x} \max_{y} f(x, y)$ to get an lower bound of (6.46) as

$$-\widetilde{G}(\eta) = \min_{f_{k,i}^{p}} \max_{\mathbf{\Delta}_{\mathcal{K}}} \sum_{k,i,p} \log\left(\mathrm{MSE}_{k,i}^{p}\right) + \eta \sum_{k,i,p} \|\boldsymbol{p}_k^{p,i}\|^2 + \eta P_c. \tag{6.48}$$

By substituting $G(\eta)$ in (6.38) with $\widetilde{G}(\eta)$, the optimization (6.38) is relaxed to

$$\begin{aligned}&\min_{\boldsymbol{P}_k^p,\eta,f_{k,i}^p}\ \max_{\boldsymbol{\Delta}_{\mathcal{K}}}\sum_{k,i,p}\log\left(\mathrm{MSE}_{k,i}^p\right)+\eta\sum_{k,i,p}\left\|\boldsymbol{p}_k^{p,i}\right\|^2+\eta P_c,\\&\quad \text{s.t. } \|\boldsymbol{P}\|_F^2\le P_{\max},\end{aligned}\tag{6.49}$$

where $\forall i\in\left\{1,\ldots,\frac{N_u}{2}\right\}, k\in\mathcal{K}, p\in\{h,v\},\eta$. The above problem is still nonconvex because the optimization variables are coupled. To tackle this nonconvexity, it is necessary to utilize some mathematical transformations. First, the term $\log\left(\mathrm{MSE}_{k,i}^p\right)$ in the objective of the optimization problem (6.49) is replaced by an auxiliary function $U_{k,i}^p\left(\alpha_{k,i}^p\right)$. Specifically, defining

$$U_{k,i}^p\left(\alpha_{k,i}^p\right)=\exp\left(\alpha_{k,i}^p-1\right)\mathrm{MSE}_{k,i}^p-\alpha_{k,i}^p,\tag{6.50}$$

it is straightforward that (also see [38])

$$\min_{\alpha_{k,i}^p} U_{k,i}^p\left(\alpha_{k,i}^p\right)=\log\left(\mathrm{MSE}_{k,i}^p\right).\tag{6.51}$$

Then based on (6.50) and (6.51), the optimization problem (6.49) can be reformulated as

$$\begin{aligned}&\min_{\boldsymbol{P}_k^p,\eta,f_{k,i}^p,\alpha_{k,i}^p}\ \max_{\boldsymbol{\Delta}_{\mathcal{K}}}\sum_{k,i,p}\exp\left(\alpha_{k,i}^p-1\right)\mathrm{MSE}_{k,i}^p-\alpha_{k,i}^p+\eta\sum_{k,i,p}\left\|\boldsymbol{p}_k^{p,i}\right\|^2+\eta P_c,\\&\quad \text{s.t. } \|\boldsymbol{P}\|_F^2\le P_{\max}.\end{aligned}\tag{6.52}$$

Moreover, substituting $\mathrm{MSE}_{k,i}^p$ in (6.44) into (6.52), the optimization problem (6.54) can be further reformulated as the following optimization problem

$$\begin{aligned}&\min_{\boldsymbol{P}_k^p,\eta,f_{k,i}^p,\alpha_{k,i}^p}\ \gamma,\\&\text{s.t.} \sum_{k,i,p} l_{k,i}^p\left\|\boldsymbol{h}_{kk,i}^p\boldsymbol{P}_k^p-\frac{1}{f_{k,i}^p}\boldsymbol{e}_i\right\|^2+\sum_{k,i,p} l_{k,i}^p\sum_{j\in\mathcal{K}\backslash k}\left\|\boldsymbol{h}_{kj,i}^p\boldsymbol{P}_j^p\right\|^2\\&\quad+\eta\sum_{k,i,p}\left\|\boldsymbol{p}_k^{p,i}\right\|^2+c\le\gamma,\ \|\boldsymbol{P}\|_F^2\le P_{\max},\end{aligned}\tag{6.53}$$

where $l_{k,i}^p$ in the first constraint is defined as follows

$$l_{k,i}^p=\exp\left(\alpha_{k,i}^p-1\right)\left|f_{k,i}^p\right|^2,\ c=\eta P_c+\sum_{k,i,p}\left(\sigma_n^2 l_{k,i}^p-\alpha_{k,i}^p\right).\tag{6.54}$$

To tackle the nonconvex optimization (6.53), which is actually an epigraph form of the problem (6.52) [39], we propose an alternating algorithm in which the problem (6.52) is transformed to a series of subproblems that can be solved efficiently in an iterative procedure. In particular, the $\boldsymbol{P}_k^p$ and $f_{k,i}^p$ related subproblems can be cast into standard SDP problems via the sign-definiteness lemma (detailed in Appendix 1) [40]. The key idea behind the proposed iterative optimization algorithm is to exploit iterative procedures to decouple the multiple optimization variables. In each iteration, some necessary mathematical transformations are used to reveal the hidden convexity. In the following, we discuss the proposed iterative optimization algorithm for the optimization problem (6.53) in detail.

6.3.2.1 Optimizing $\alpha_{k,i}^p$

When $\boldsymbol{P}_k^p$ and $f_{k,i}^p$ are fixed, the function (6.50) is convex with respect to $\alpha_{k,i}^p$, so by setting the derivative of $U_{k,i}^p(\alpha_{k,i}^p)$ equal to zero, the optimal value of $\alpha_{k,i}^p$ can be obtained as

$$\alpha_{k,i}^{p\star} = -\log\left(\mathrm{MSE}_{k,i}^p\right) + 1. \tag{6.55}$$

Actually, we can obtain the minimum of $U_{k,i}^p(\alpha_{k,i}^p)$ as $\log\left(\mathrm{MSE}_{k,i}^p\right)$ by substituting (6.55) into (6.50), which is consistent with (6.51). Because of the concavity of the log function [38], the optimization (6.55) is a standard convex problem clearly.

6.3.2.2 Optimizing $\boldsymbol{P}_k^p$

When $\alpha_{k,i}^p$ and $f_{k,i}^p$ are fixed, the optimization (6.53) becomes the problem of $\min_{\boldsymbol{P}_k^p} \gamma$. In this case, its constraints can be transferred into a simplified form by introducing the auxiliary variables $b_{kj,i}^p$ and $d_k^{p,i}$, and then the optimization problem can be rewritten as

$$\begin{aligned}
&\min_{\boldsymbol{P}_k^p, b_{kj,i}^p, d_k^{p,i}} \quad \gamma, \\
&\text{s.t.} \sqrt{l_{k,i}^p}\left\| \boldsymbol{h}_{kk,i}^p \boldsymbol{P}_k^p - \frac{1}{f_{k,i}^p}\boldsymbol{e}_i \right\| \le b_{kk,i}^p, \\
&\quad \|\boldsymbol{\Delta}_{kk,i}^p\| \le \tau_{kk,i}^p, \sqrt{l_{k,i}^p}\|\boldsymbol{h}_{kj,i}^p \boldsymbol{P}_j^p\| \le b_{kj,i}^p, \\
&\quad \|\boldsymbol{\Delta}_{kj,i}^p\| \le \tau_{kj,i}^p, \sqrt{\eta}\|\boldsymbol{p}_k^{p,i}\| \le d_k^{p,i}, \\
&\quad \sum_{k,i,p}\Big(\sum_{j\in\mathcal{K}} (b_{kj,i}^p)^2 + (d_k^{p,i})^2\Big) + c \le \gamma, \\
&\quad \|\boldsymbol{P}\|_F^2 \le P_{\max}.
\end{aligned} \tag{6.56}$$

According to Schur complement lemma [41], the first four sets of constraints in (6.56) are re-expressed as the equivalent semi-infinite linear matrix inequality (LMI) with $\forall k, j \in \mathcal{K}, j \neq k, i \in \left\{1, \ldots, \frac{N_u}{2}\right\}, p \in \{h, v\}, \eta$, which is presented in (6.57). By defining the variables in (6.58), which is given on the next page, the semi-infinite constraints (6.57) can be further transferred into the following equivalent finite linear matrix inequality LMIs

$$\begin{aligned}
&\forall k, j \in \mathcal{K},\ j \neq k,\ i \in \left\{1, \ldots, \tfrac{N_u}{2}\right\},\ p \in \{h, v\}, \eta: \\
&\begin{bmatrix} \frac{b^p_{kk,i}}{\sqrt{l^p_{k,i}}} & \left(\widehat{\boldsymbol{h}}^p_{kk,i} + \boldsymbol{\Delta}^p_{kk,i}\right)\boldsymbol{P}^p_k - \frac{1}{f^p_{k,i}}\boldsymbol{e}_i \\ \left(\boldsymbol{P}^p_k\right)^{\mathrm{H}}\left(\widehat{\boldsymbol{h}}^p_{kk,i} + \boldsymbol{\Delta}^p_{kk,i}\right)^{\mathrm{H}} - \frac{1}{\left(f^p_{k,i}\right)^*}\boldsymbol{e}^{\mathrm{H}}_i & \frac{b^p_{kk,i}}{\sqrt{l^p_{k,i}}}\boldsymbol{I}_{\frac{N_u}{2}} \end{bmatrix} \succeq \boldsymbol{0},\ \|\boldsymbol{\Delta}^p_{kk,i}\| \leq \tau^p_{kk,i}, \\
&\begin{bmatrix} \frac{b^p_{kj,i}}{\sqrt{l^p_{k,i}}} & \left(\widehat{\boldsymbol{h}}^p_{kj,i} + \boldsymbol{\Delta}^p_{kj,i}\right)\boldsymbol{P}^p_j \\ \left(\boldsymbol{P}^p_j\right)^{\mathrm{H}}\left(\widehat{\boldsymbol{h}}^p_{kj,i} + \boldsymbol{\Delta}^p_{kj,i}\right)^{\mathrm{H}} & \frac{b^p_{kj,i}}{\sqrt{l^p_{k,i}}}\boldsymbol{I}_{\frac{N_u}{2}} \end{bmatrix} \succeq \boldsymbol{0},\ \|\boldsymbol{\Delta}^p_{kj,i}\| \leq \tau^p_{kj,i}.
\end{aligned} \tag{6.57}$$

$$\begin{aligned}
&\forall k, j \in \mathcal{K},\ j \neq k,\ i \in \left\{1, \ldots, \tfrac{N_u}{2}\right\},\ p \in \{h, v\}, \eta: \\
&\quad \boldsymbol{X} = -[1\ \boldsymbol{0}] \in \mathbb{R}^{1\times\left(1+\frac{N_u}{2}\right)},\ \boldsymbol{Y}^p_k = \left[\boldsymbol{0}\ \boldsymbol{P}^p_k\right] \in \mathbb{C}^{\frac{L}{2}\times\left(1+\frac{N_u}{2}\right)},\ \boldsymbol{Q}^p_{kj,i} = \boldsymbol{\Delta}^p_{kj,i} \in \mathbb{C}^{1\times\frac{L}{2}}, \\
&\quad \boldsymbol{Z}^p_{kk,i} = \begin{bmatrix} \frac{b^p_{kk,i}}{\sqrt{l^p_{k,i}}} & \widehat{\boldsymbol{h}}^p_{kk,i}\boldsymbol{P}^p_k - \frac{1}{f^p_{k,i}}\boldsymbol{e}_i \\ \left(\boldsymbol{P}^p_k\right)^{\mathrm{H}}\left(\widehat{\boldsymbol{h}}^p_{kk,i}\right)^{\mathrm{H}} - \frac{1}{\left(f^p_{k,i}\right)^*}\boldsymbol{e}^{\mathrm{H}}_i & \frac{b^p_{kk,i}}{\sqrt{l^p_{k,i}}}\boldsymbol{I}_{\frac{N_u}{2}} \end{bmatrix} \in \mathbb{C}^{\left(1+\frac{N_u}{2}\right)\times\left(1+\frac{N_u}{2}\right)}, \\
&\quad \boldsymbol{Z}^p_{kj,i} = \begin{bmatrix} \frac{b^p_{kj,i}}{\sqrt{l^p_{k,i}}} & \widehat{\boldsymbol{h}}^p_{kj,i}\boldsymbol{P}^p_j \\ \left(\boldsymbol{P}^p_j\right)^{\mathrm{H}}\left(\widehat{\boldsymbol{h}}^p_{kj,i}\right)^{\mathrm{H}} & \frac{b^p_{kj,i}}{\sqrt{l^p_{k,i}}}\boldsymbol{I}_{\frac{N_u}{2}} \end{bmatrix} \in \mathbb{C}^{\left(1+\frac{N_u}{2}\right)\times\left(1+\frac{N_u}{2}\right)},
\end{aligned} \tag{6.58}$$

$$\begin{aligned}
&\forall k, j \in \mathcal{K},\ j \neq k,\ i \in \left\{1, \ldots, \tfrac{N_u}{2}\right\},\ p \in \{h, v\}, \eta: \\
&\boldsymbol{Z}^p_{kk,i} \succeq \boldsymbol{X}^{\mathrm{H}}\boldsymbol{Q}^p_{kk,i}\boldsymbol{Y}^p_k + \left(\boldsymbol{Y}^p_k\right)^{\mathrm{H}}\left(\boldsymbol{Q}^p_{kk,i}\right)^{\mathrm{H}}\boldsymbol{X}, \\
&\boldsymbol{Z}^p_{kj,i} \succeq \boldsymbol{X}^{\mathrm{H}}\boldsymbol{Q}^p_{kj,i}\boldsymbol{Y}^p_j + \left(\boldsymbol{Y}^p_j\right)^{\mathrm{H}}\left(\boldsymbol{Q}^p_{kj,i}\right)^{\mathrm{H}}\boldsymbol{X}.
\end{aligned} \tag{6.59}$$

Naturally, by utilizing Lemma 6.1, the optimization problem (6.56) can be rewritten as the optimization (6.60), which is also presented at the bottom of the next page. It is obvious that the optimization problem (6.60) is a standard SDP problem. Therefore the optimization problem (6.60) can be solved effectively by using inter-point algorithms.

$$
\begin{aligned}
\min_{\boldsymbol{P}_k^p, b_{kj,i}^p, d_k^{p,i}} \quad & \gamma, \\
\text{s.t.} \quad & \forall k, j \in \mathcal{K},\ j \neq k,\ i \in \left\{1, \ldots, \tfrac{N_u}{2}\right\},\ p \in \{h, v\}, \eta: \\
& \begin{bmatrix}
\frac{b_{kk,i}^p}{\sqrt{l_{k,i}^p}} - \lambda_{kk,i}^p & \widehat{\boldsymbol{h}}_{kk,i}^p \boldsymbol{P}_k^p - \frac{1}{f_{k,i}^p} \boldsymbol{e}_i & \boldsymbol{0} \\
(\boldsymbol{P}_k^p)^{\mathrm{H}} (\widehat{\boldsymbol{h}}_{kk,i}^p)^{\mathrm{H}} - \frac{1}{(f_{k,i}^p)^*} \boldsymbol{e}_i^{\mathrm{H}} & \frac{b_{kk,i}^p}{\sqrt{l_{k,i}^p}} \boldsymbol{I}_{\frac{N_u}{2}} & -\tau_{kk,i}^p (\boldsymbol{P}_k^p)^{\mathrm{H}} \\
\boldsymbol{0} & -\tau_{kk,i}^p \boldsymbol{P}_k^p & \lambda_{kk,i}^p \boldsymbol{I}_{\frac{L}{2}}
\end{bmatrix} \succeq \boldsymbol{0}, \\
& \begin{bmatrix}
\frac{b_{kj,i}^p}{\sqrt{l_{k,i}^p}} - \lambda_{kj,i}^p & \widehat{\boldsymbol{h}}_{kj,i}^p \boldsymbol{P}_j^p & \boldsymbol{0} \\
(\boldsymbol{P}_j^p)^{\mathrm{H}} (\widehat{\boldsymbol{h}}_{kj,i}^p)^{\mathrm{H}} & \frac{b_{kj,i}^p}{\sqrt{l_{k,i}^p}} \boldsymbol{I}_{\frac{N_u}{2}} & -\tau_{kj,i}^p (\boldsymbol{P}_j^p)^{\mathrm{H}} \\
\boldsymbol{0} & -\tau_{kj,i}^p \boldsymbol{P}_j^p & \lambda_{kj,i}^p \boldsymbol{I}_{\frac{L}{2}}
\end{bmatrix} \succeq \boldsymbol{0}, \\
& \lambda_{kk,i}^p > 0,\ \lambda_{kj,i}^p > 0,\ \sqrt{\eta} \| \boldsymbol{p}_k^{p,i} \| \leq d_k^{p,i}, \\
& \sum_{k,i,p} \Big(\sum_{j \in \mathcal{K}} \left(b_{kj,i}^p\right)^2 + \left(d_k^{p,i}\right)^2 \Big) + c \leq \gamma,\ \|\boldsymbol{P}\|_F^2 \leq P_{\max}.
\end{aligned}
\tag{6.60}
$$

6.3.2.3 Optimizing $f_{k,i}^p$

Given $\boldsymbol{P}_k^p$ and $\alpha_{k,i}^p$, the optimization problem (6.53) can be expressed as

$$
\begin{aligned}
\min_{f_{k,i}^p} \quad & \gamma, \\
\text{s.t.} \quad & \max_{\boldsymbol{\Delta}_{\mathcal{K}}} \sum_{k,i,p} \exp\left(\alpha_{k,i}^p - 1\right) \mathrm{MSE}_{k,i}^p \leq \gamma - g,
\end{aligned}
\tag{6.61}
$$

where g on the righthand side of the inequality is defined as

$$
g = \sum_{k,i,p} -\alpha_{k,i}^p + \eta \left\| \boldsymbol{p}_k^{p,i} \right\|^2 + \eta P_c. \tag{6.62}
$$

By substituting $\mathrm{MSE}_{k,i}^p$ of (6.44) into (6.61) and defining the auxiliary variables $q_{kj,i}^p$ that satisfy

$$
\begin{aligned}
& \| f_{k,i}^p \boldsymbol{h}_{kk,i}^p \boldsymbol{P}_k^p - \boldsymbol{e}_i \| \leq q_{kk,i}^p,\ \| f_{k,i}^p \boldsymbol{h}_{kj,i}^p \boldsymbol{P}_j^p \| \leq q_{kj,i}^p, \\
& \sum_{k,i,p} \exp\left(\alpha_{k,i}^p - 1\right) \Big(\sum_{j \in \mathcal{K}} \left(q_{kj,i}^p\right)^2 + \sigma_n^2 |f_{k,i}^p|^2 \Big) \leq \gamma - g,
\end{aligned}
\tag{6.63}
$$

the optimization problem (6.61) becomes

$$\begin{aligned}
&\min_{f^p_{k,i}, q^p_{kj,i}} \quad \gamma, \\
&\text{s.t.} \ \|f^p_{k,i}\boldsymbol{h}^p_{kk,i}\boldsymbol{P}^p_k - \boldsymbol{e}_i\| \le q^p_{kk,i}, \\
&\qquad \sum_{k,i,p} \exp\left(\alpha^p_{k,i} - 1\right)\Big(\sum_{j\in\mathcal{K}} \left(q^p_{kj,i}\right)^2 + \sigma_n^2 |f^p_{k,i}|^2\Big) \le \gamma - g \\
&\qquad \|\boldsymbol{\Delta}^p_{kk,i}\| \le \tau^p_{kk,i}, \ \|f^p_{k,i}\boldsymbol{h}^p_{kj,i}\boldsymbol{P}^p_j\| \le q^p_{kj,i}, \\
&\qquad \|\boldsymbol{\Delta}^p_{kj,i}\| \le \tau^p_{kj,i}.
\end{aligned} \tag{6.64}$$

$$\begin{aligned}
&\begin{bmatrix} q^p_{kk,i} & f^p_{k,i}\left(\widehat{\boldsymbol{h}}^p_{kk,i} + \boldsymbol{\Delta}^p_{kk,i}\right)\boldsymbol{P}^p_k - \boldsymbol{e}_i \\ \left(f^p_{k,i}\right)^* \left(\boldsymbol{P}^p_k\right)^{\mathrm{H}} \left(\widehat{\boldsymbol{h}}^p_{kk,i} + \boldsymbol{\Delta}^p_{kk,i}\right)^{\mathrm{H}} - \boldsymbol{e}_i^{\mathrm{H}} & q^p_{kk,i}\boldsymbol{I}_{\frac{N_u}{2}} \end{bmatrix} \succeq \boldsymbol{0}, \\
&\begin{bmatrix} q^p_{kj,i} & f^p_{k,i}\left(\widehat{\boldsymbol{h}}^p_{kj,i} + \boldsymbol{\Delta}^p_{kj,i}\right)\boldsymbol{P}^p_j \\ \left(f^p_{k,i}\right)^* \left(\boldsymbol{P}^p_j\right)^{\mathrm{H}} \left(\widehat{\boldsymbol{h}}^p_{kj,i} + \boldsymbol{\Delta}^p_{kj,i}\right)^{\mathrm{H}} & q^p_{kj,i}\boldsymbol{I}_{\frac{N_u}{2}} \end{bmatrix} \succeq \boldsymbol{0}, \\
&\left\|\boldsymbol{\Delta}^p_{kk,i}\right\| \le \tau^p_{kk,i}, \quad \left\|\boldsymbol{\Delta}^p_{kj,i}\right\| \le \tau^p_{kj,i}.
\end{aligned} \tag{6.65}$$

$$\begin{aligned}
&\boldsymbol{X} = -[1\ \boldsymbol{0}] \in \mathbb{R}^{1\times\left(1+\frac{N_u}{2}\right)}, \boldsymbol{Y}^p_{kj,i} = \left[\boldsymbol{0}\ f^p_{k,i}\boldsymbol{P}^p_j\right] \in \mathbb{C}^{\frac{L}{2}\times\left(1+\frac{N_u}{2}\right)}, \boldsymbol{Q}^p_{kj,i} = \boldsymbol{\Delta}^p_{kj,i} \in \mathbb{C}^{1\times\frac{L}{2}}, \\
&\boldsymbol{Z}^p_{kk,i} = \begin{bmatrix} q^p_{kk,i} & f^p_{k,i}\widehat{\boldsymbol{h}}^p_{kk,i}\boldsymbol{P}^p_k - \boldsymbol{e}_i \\ \left(f^p_{k,i}\right)^*\left(\boldsymbol{P}^p_k\right)^{\mathrm{H}}\left(\widehat{\boldsymbol{h}}^p_{kk,i}\right)^{\mathrm{H}} - \boldsymbol{e}_i^{\mathrm{H}} & q^p_{kk,i}\boldsymbol{I}_{\frac{N_u}{2}} \end{bmatrix} \in \mathbb{C}^{\left(1+\frac{N_u}{2}\right)\times\left(1+\frac{N_u}{2}\right)}, \\
&\boldsymbol{Z}^p_{kj,i} = \begin{bmatrix} q^p_{kj,i} & f^p_{k,i}\widehat{\boldsymbol{h}}^p_{kj,i}\boldsymbol{P}^p_j \\ \left(f^p_{k,i}\right)^*\left(\boldsymbol{P}^p_j\right)^{\mathrm{H}}\left(\widehat{\boldsymbol{h}}^p_{kj,i}\right)^{\mathrm{H}} & q^p_{kj,i}\boldsymbol{I}_{\frac{N_u}{2}} \end{bmatrix} \in \mathbb{C}^{\left(1+\frac{N_u}{2}\right)\times\left(1+\frac{N_u}{2}\right)}, \ j \ne k,
\end{aligned} \tag{6.66}$$

Similar to the optimizing of $\boldsymbol{P}^p_k$, by utilizing Schur complement lemma, the constraints in (6.64) can also be transformed into the semi-infinite LMIs with $\forall k, j \in \mathcal{K}$, $j \ne k$, $i \in \left\{1, \ldots, \frac{N_u}{2}\right\}$, $p \in \{h, v\}, \eta$. The corresponding expressions are given in (6.65) on the next page. For $\forall k, j \in \mathcal{K}, i \in \left\{1, \ldots, \frac{N_u}{2}\right\}$, $p \in \{h, v\}, \eta$, by defining the variables in (6.66), the two sets of semi-infinite LMIs in (6.65) can be transformed into the finite LMIs given in the following equations

$$\begin{aligned}
&\forall k, j \in \mathcal{K}, \ j \ne k, \ i \in \left\{1, \ldots, \tfrac{N_u}{2}\right\}, \ p \in \{h, v\}, \eta: \\
&\boldsymbol{Z}^p_{kk,i} \succeq \boldsymbol{X}^{\mathrm{H}}\boldsymbol{Q}^p_{kk,i}\boldsymbol{Y}^p_{kk,i} + \left(\boldsymbol{Y}^p_{kk,i}\right)^{\mathrm{H}}\left(\boldsymbol{Q}^p_{kk,i}\right)^{\mathrm{H}}\boldsymbol{X}, \\
&\boldsymbol{Z}^p_{kj,i} \succeq \boldsymbol{X}^{\mathrm{H}}\boldsymbol{Q}^p_{kj,i}\boldsymbol{Y}^p_{kj,i} + \left(\boldsymbol{Y}^p_{kj,i}\right)^{\mathrm{H}}\left(\boldsymbol{Q}^p_{kj,i}\right)^{\mathrm{H}}\boldsymbol{X}.
\end{aligned} \tag{6.67}$$

Exploiting Lemma 6.1, the SDP optimization problem for $f_{k,i}^{p}$ is given by

$$\min_{f_{k,i}^{p}, q_{kj,i}^{p}} \gamma,$$
$$\text{s.t.} \quad \forall k, j \in \mathcal{K},\ j \neq k,\ i \in \left\{1, \ldots, \tfrac{N_u}{2}\right\},\ p \in \{h, v\},\ \eta :$$
$$\begin{bmatrix} q_{kk,i}^{p} - \beta_{kk,i}^{p} & f_{k,i}^{p} \widehat{\boldsymbol{h}}_{kk,i}^{p} \boldsymbol{P}_{k}^{p} - \boldsymbol{e}_i & \boldsymbol{0} \\ (f_{k,i}^{p})^{*} (\boldsymbol{P}_{k}^{p})^{\mathrm{H}} (\widehat{\boldsymbol{h}}_{kk,i}^{p})^{\mathrm{H}} - \boldsymbol{e}_i^{\mathrm{H}} & q_{kk,i}^{p} \boldsymbol{I}_{\frac{N_u}{2}} & -\tau_{kk,i}^{p} (f_{k,i}^{p})^{*} (\boldsymbol{P}_{k}^{p})^{\mathrm{H}} \\ \boldsymbol{0} & -\tau_{kk,i}^{p} f_{k,i}^{p} \boldsymbol{P}_{k}^{p} & \beta_{kk,i}^{p} \boldsymbol{I}_{\frac{L}{2}} \end{bmatrix} \succeq \boldsymbol{0},$$
$$\begin{bmatrix} q_{kj,i}^{p} - \beta_{kj,i}^{p} & f_{k,i}^{p} \widehat{\boldsymbol{h}}_{kj,i}^{p} \boldsymbol{P}_{j}^{p} & \boldsymbol{0} \\ (f_{k,i}^{p})^{*} (\boldsymbol{P}_{j}^{p})^{\mathrm{H}} (\widehat{\boldsymbol{h}}_{kj,i}^{p})^{\mathrm{H}} & q_{kj,i}^{p} \boldsymbol{I}_{\frac{N_u}{2}} & -\tau_{kj,i}^{p} (f_{k,i}^{p})^{*} (\boldsymbol{P}_{j}^{p})^{\mathrm{H}} \\ \boldsymbol{0} & -\tau_{kj,i}^{p} f_{k,i}^{p} \boldsymbol{P}_{j}^{p} & \beta_{kj,i}^{p} \boldsymbol{I}_{\frac{L}{2}} \end{bmatrix} \succeq \boldsymbol{0},$$
$$\beta_{kk,i}^{p} > 0,\ \beta_{kj,i}^{p} > 0,\ \sum_{k,i,p} \exp\left(\alpha_{k,i}^{p} - 1\right)\Big(\sum_{j \in \mathcal{K}} \left(q_{kj,i}^{p}\right)^2 + \sigma_n^2 |f_{k,i}^{p}|^2 \Big) \leq \gamma - g. \tag{6.68}$$

In a nutshell, when the value of $\eta$is fixed, in order to solve the optimization problem (6.53) with respect to the variables $\left\{\alpha_{k,i}^{p}, \boldsymbol{P}_{k}^{p}, f_{k,i}^{p} : \forall i \in \left\{1, \ldots, \frac{N_u}{2}\right\}, k \in \mathcal{K}, p \in \{h, v\}\right\}$ effectively, an iterative optimization algorithm can be implemented among the three sub-optimization problems, i.e., the convex problem (6.55), the SDP (6.60) and the SDP (6.68), iteratively.

6.3.2.4 Optimizing η

Incorporating the bisection method to optimize η with the above iterative optimization procedure, the optimal $\underline{\eta}^{\star}$ for the optimization problem (6.49), which is equivalent to the problem (6.52), is obtained as

$$\max_{\boldsymbol{P}_{k}^{p}} \widetilde{G}\left(\underline{\eta}^{\star}\right) = 0, \tag{6.69}$$

where $\widetilde{G}(\eta)$ is given in (6.48). Note that $\underline{\eta}^{\star}$ is a lower bound to our original worst-case EE optimization problem (6.34). In fact, by substituting $\underline{\eta}^{\star}$ into the optimization problem (6.38) and utilizing the fact that $\widetilde{G}(\eta)$ is a lower bound of the objective function of the problem (6.47), we have

$$G_{org}\left(\underline{\eta}^{\star}\right) = \max_{\boldsymbol{P}_{k}^{p}} G\left(\underline{\eta}^{\star}\right) \geq \max_{\boldsymbol{P}_{k}^{p}} \widetilde{G}\left(\underline{\eta}^{\star}\right) = 0. \tag{6.70}$$

Since the globally optimal EE solution $\eta^{\star}$ for solving the optimization (6.38) satisfies $G_{org}(\eta^{\star}) = 0$, from (6.70), we have $G_{org}\left(\underline{\eta}^{\star}\right) \geq G_{org}(\eta^{\star})$. Further considering that $G_{org}(\eta)$ is a decreasing function of η, it can be concluded that $\eta^{\star} \geq \underline{\eta}^{\star}$. In other words, $\underline{\eta}^{\star}$ is a lower bound of the optimal $\eta^{\star}$. Moreover, because the optimization problems (6.38) and (6.34) are equivalent to each other, $\underline{\eta}^{\star}$ is a lower bound of the

Algorithm 6.1 Iterative procedure for obtaining optimal linear precoding matrices

Require: The initial linear precoding $\boldsymbol{P}_k^{p(0)}$, the initial equalizer $f_{k,i}^{p(0)}$, the initial lower bound $\eta_{\min} = 0$, and the initial upper bound $\eta_{\max} = \frac{\sum_{i,k,p} R_{k,i}^{p\max}}{P_C}$;

Set the inner optimization accuracy ϵ_{inn} (or the maximum number of inner iterations I_{inn});

Set the outer bisection search accuracy ϵ_{out} (or the maximum number of outer iterations I_{out});

Set the outer iteration index $u = 0$;

repeat

  $\eta = \frac{\eta_{\max}+\eta_{\min}}{2}$;

  Set the inner iteration index $t = 0$;

  repeat

    Calculate $\alpha_{k,i}^{p(t+1)}$ by solving (6.55) for given η, $\boldsymbol{P}_k^{p(t)}$, and $f_{k,i}^{p(t)}$;

    Calculate $\boldsymbol{P}_k^{p(t+1)}$ by solving (6.60) for given η, $f_{k,i}^{p(t)}$, and $\alpha_{k,i}^{p(t+1)}$;

    Calculate $f_{k,i}^{p(t+1)}$ by solving (6.68) for given η, $\alpha_{k,i}^{p(t+1)}$, and $\boldsymbol{P}_k^{p(t+1)}$;

    $t = t + 1$;

  until $\|\boldsymbol{P}_k^{p(t+1)} - \boldsymbol{P}_k^{p(t)}\| < \epsilon_{\text{inn}}, k \in \mathcal{K}, p \in \{h, v\}$ (or $t == I_{\text{inn}}$);

  if $\gamma > 0$ **then**

    $\eta_{\min} = \eta$;

  else

    $\eta_{\max} = \eta$;

  end if

  $u = u + 1$;

until $\eta_{\max} - \eta_{\min} < \epsilon_{\text{out}}$ (or $u == I_{\text{out}}$);

$\boldsymbol{P}_k^{p\star} = \boldsymbol{P}_k^{p(t)}, k \in \mathcal{K},\ p \in \{h, v\}$;

Ensure: The optimal linear precoding $\boldsymbol{P}_k^{p\star}, k \in \mathcal{K},\ p \in \{h, v\}$.

optimal worst-case EE, which is our original optimization objective. At the end, the proposed iterative algorithm that computes the optimal linear precoding matrices $\boldsymbol{P}_k^{p\star}, \forall k \in \mathcal{K},\ p \in \{h, v\}$ is summarized in Algorithm 6.1.

6.4 Convergence and Complexity Analysis

Theorem 6.1 *Algorithm 6.1 converges to a stationary point of problem (6.52).*

Proof See Appendix 2.

It is readily observed that the computational complexity of our proposed Algorithm 6.1 mainly comes from solving two SDP problems in terms of $\boldsymbol{P}_k^p$ and $f_{k,i}^p, \forall i, k, p$. Referring to [39], the computational complexity of the standard SDP programming is $\mathcal{O}\big(M_{\text{sdp}}N_{\text{sdp}}^{3.5} + M_{\text{sdp}}^2 N_{\text{sdp}}^{2.5} + M_{\text{sdp}}^3 N_{\text{sdp}}^{0.5}\big) \log\left(\frac{1}{\epsilon}\right)$, where M_{sdp} denotes the number of semi-definite cone constraints and N_{sdp} is the dimension of semi-definite cones, while ϵ denotes the accuracy imposed to solve the SDP problem. In particular, both the optimizing $\boldsymbol{P}_k^p$ and $f_{k,i}^p$ have the same number of semi-definite cone constraints $M_{\text{sdp}} = K^2 N_u$ and the same semi-definite cone dimension

$N_{\text{sdp}} = 1 + \frac{N_u}{2} + \frac{L}{2}$. By neglecting the constant term and the low-order term, the complexity order of the proposed Algorithm 6.1 is given by

$$\mathsf{O}_1 = (I_{\text{inn}} I_{\text{out}}) \mathsf{O}\left(K^2 N_u \left(1 + \frac{N_u}{2} + \frac{L}{2}\right)^{3.5} + (K^2 N_u)^2 \left(1 + \frac{N_u}{2} + \frac{L}{2}\right)^{2.5} + (K^2 N_u)^3 \left(1 + \frac{N_u}{2} + \frac{L}{2}\right)^{0.5} \right) \log\left(\frac{1}{\epsilon}\right). \tag{6.71}$$

6.5 Numerical Results and Analysis

For the spatial covariance matrices of the polarized MIMO channels, the widely used one-ring model is adopted in the simulation. According to the one-ring model, the mth row and nth column element of $\boldsymbol{R}_k$ for the MUs in the kth group is calculated based on the following equation [42]

$$[\boldsymbol{R}_k]_{m,n} = \frac{1}{2\Delta_k^a} \int_{-\Delta_k^a}^{\Delta_k^a} e^{-\mathrm{j}\pi \lambda_0^{-1} \boldsymbol{\Omega}(\alpha+\theta_k)(\boldsymbol{r}_m - \boldsymbol{r}_n)} d\alpha, \tag{6.72}$$

where λ_0 is the transmit RF signal wavelength, θ_k is the azimuth direction of the kth MU group. In addition, the vector $\boldsymbol{\Omega}(\alpha) = [\cos(\alpha)\ \ \sin(\alpha)]$ denotes the RF signal departure direction vector with angle α, while $\boldsymbol{r}_m$ denotes the location vector of the mth antenna, and $\Delta_k^a \approx \tan^{-1}(\varpi_k/\varrho_k)$ is the angular spread of the departure RF signal to the kth MU group, where ϱ_k represents the distance between the BS and the kth MU group, ϖ_k is the radius of the scattering ring. Without loss of generality, it is assumed that $\Delta_1^a = \cdots = \Delta_K^a = \frac{4\pi}{180}$, and the azimuth angle of the kth MU group satisfies $\theta_k = -\frac{\pi}{6} + \frac{\pi}{3}(k-1), \forall k \in \mathcal{K}$. The simulated dual-polarized multiuser MIMO system contains one BS equipped with N_t polarized antennas and $K = 2$ MU groups each with $N_u = 2$ MUs. The spacing between adjacent antenna elements of a dual-polarized array is half of the transmit signal wavelength. All the channel coefficients, namely, the elements of all $\boldsymbol{G}_k$, are assumed to be i.i.d. complex Gaussian distributed with zero mean and variance $\sigma^2 = 1$. The cross-polar discrimination parameter is set that $\chi = 0$ and the power of all the received AWGNs is $\sigma_n^2 = 1$. The default CSI error thresholds are set to $\sigma_k = \sigma = 0.1, \forall k \in \mathcal{K}$, and the circuit power consumption P_c is chosen to be a constant [32].

In our simulation, the widely used Matlab toolbox CVX [41] is adopted to solve the convex optimization problem (6.55) and the two SDP problems (6.60) and (6.68). The non-robust optimization design in which the estimated CSI is taken as the perfect CSI is simulated as a benchmark. All of the simulation curves in the following figures are averaged over 500 independent Monte Carlo trails. Theorem 6.1 proves the convergence of our proposed Algorithm 6.1 for precoding optimization. We now verify the convergence of the proposed algorithm via the simulation results when

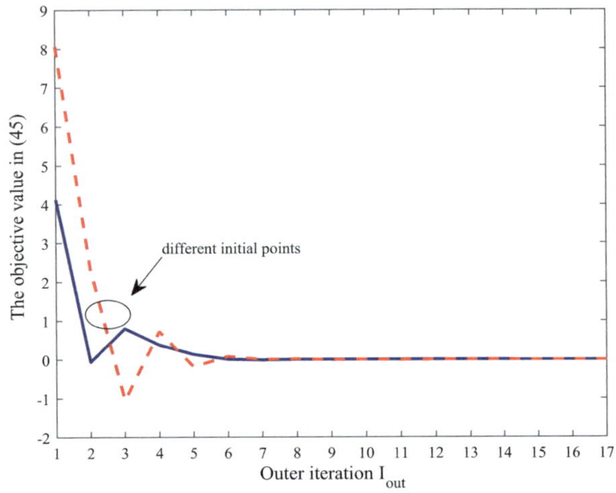

Fig. 6.1 Convergence of the outer iteration procedure of Algorithm 6.1 for different initial precoding matrices, given $N_t = 20$, $N_u = 2$, $K = 2$, $P_{\max} = 0\,\text{dBW}$, and $P_c = 0\,\text{dBW}$

$N_t = 20$, $N_u = 2$, $K = 2$, and $P_c = 0$ dBW. More specifically, Fig. 6.1 shows the convergence performance of the outer iterative process in Algorithm 6.1 with different initial values of $\boldsymbol{P}_k^{p(0)}$, which satisfy the power constraint in (6.34) with equality. Observe from Fig. 6.1 that the outer loop, namely, the bisection scheme, is guaranteed to converge with no more than 10 iterations. It can also be found that the value of the objective function for the optimization problem (6.52) is equal to zero after the convergence of Algorithm 6.1.

Based on the previous discussions in Sect. 6.3.1, the value of r involved in optimizing the preprocessing matrix $\boldsymbol{B}$ is closely related to the final system EE performance of the considered dual-polarized multiuser MIMO communication system. We now investigate in depth the impacts of r and the total transmit power $P_{\max}$ on our proposed robust design as well as on the non-robust design. Under the conditions of $N_t = 20$, $N_u = 2$, $K = 2$ and $P_c = 0\,\text{dBW}$, Fig. 6.2 shows the system EE performance of the two designs versus $P_{\max}$ for two different values of r. In addition, the conventional precoding design, i.e., directly designing $\boldsymbol{V}_k$ by Algorithm 6.1, is added as a benchmark. Noting that the worst-case system EE increases with $P_{\max}$. It can be seen that better EE performance is achieved with a higher value of r, i.e., $r = \min_{\forall k \in \mathcal{K}} r_k = 4$, and is also close to that of the conventional design. According to the complexity analysis of Algorithm 6.1, the complexity of the conventional design is in the same form as Eq. (6.71), where L is replaced by N_t, so our proposed design can significantly reduce complexity with affordable performance degradation. Moreover, Fig. 6.2 demonstrates that the proposed robust design outperforms the non-robust counterpart significantly.

Figure 6.3 shows the system EE performance of the two designs with different $P_{\max}$ shows the system EE performance of the two designs with different N_t. As expected, employing more BS antennas improves the worst-case system EE performance. From Fig. 6.3 it can be seen that the proposed robust design is better than the non-robust counterpart.

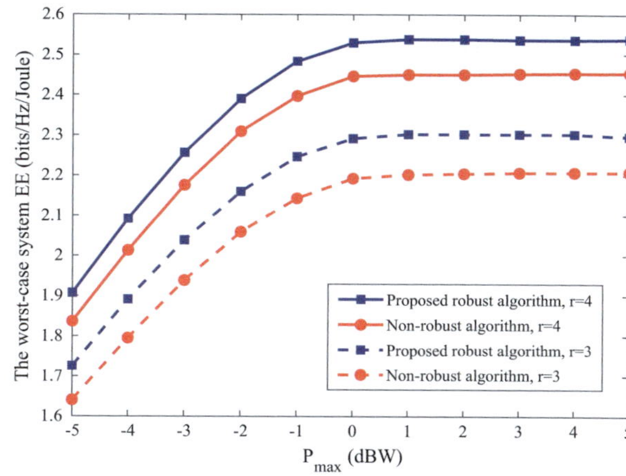

Fig. 6.2 The achievable worst-case system EE performance of the two designs as the functions of the total transmit power $P_{\max}$ for two different values of r, given $N_t = 20$, $N_u = 2$, $K = 2$, and $P_c = 0\,\text{dBW}$

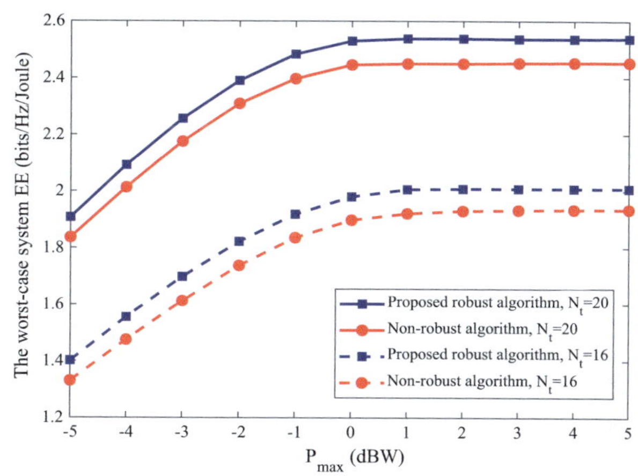

Fig. 6.3 The achievable worst-case system EE performance of the two designs as the functions of the total transmit power $P_{\max}$ for two different BS antenna numbers of $N_t = 20$ and $N_t = 16$, given $N_u = 2$, $K = 2$, and $P_c = 0\,\text{dBW}$

Figures 6.4 and 6.5 show the influence of the circuit power consumption P_c on the system EE performance. It can be found that the worst-case system EE decreases as the increase of P_c. Similar to Figs. 6.3, 6.4 shows that the system achieves better performance with more BS antennas. Furthermore, the impact of the CSI error threshold σ on the worst-case system EE is illustrated in Fig. 6.5, which shows that σ has an adverse influence on EE performance. The results of Figs. 6.4 and 6.5 again confirm the advantages of our robust design.

Moreover, when $N_t = 20$, $N_u = 2$, $K = 2$, $P_{\max} = 0\,\text{dBW}$, and $P_c = 0\,\text{dBW}$, Fig. 6.6 demonstrates the effect of the norm-bounded CSI error threshold σ on the achievable worst-case system EE performance. It can be discovered that similar to Fig. 6.5, the results of Fig. 6.6 demonstrate that the proposed robust design always outperforms the non-robust counterpart algorithm. Moreover, the EE performance becomes better when channel error σ decreases.

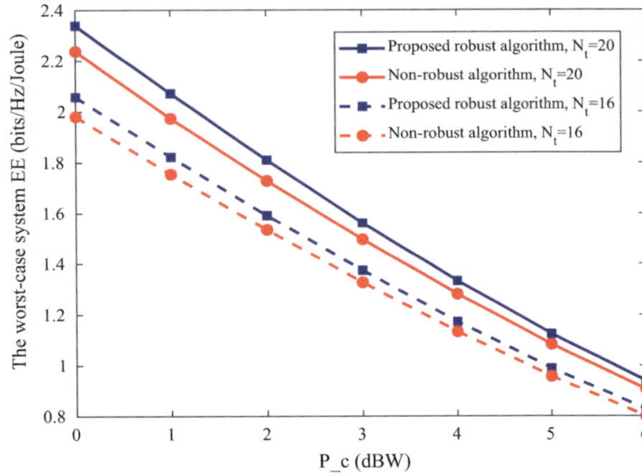

Fig. 6.4 The achievable worst-case system EE performance of the two designs as the functions of the circuit power consumption P_c for two different BS antenna numbers of $N_t = 20$ and $N_t = 16$, given $N_u = 2$, $K = 2$, and $P_{\max} = 0\,\text{dBW}$

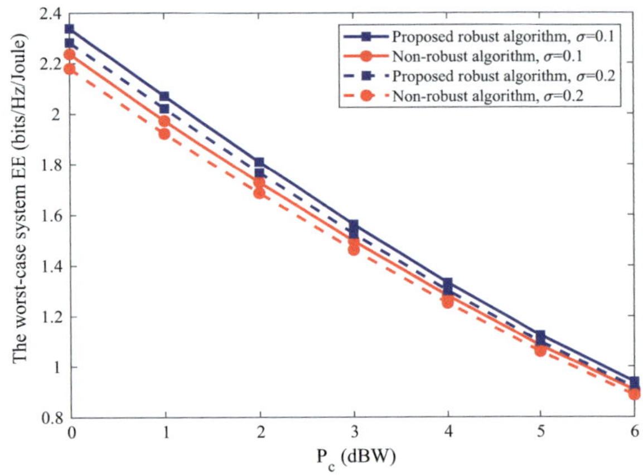

Fig. 6.5 The achievable worst-case system EE performance of the two designs as the functions of the circuit power consumption P_c for two different CSI error thresholds, given $N_t = 20$, $N_u = 2$, $K = 2$, and $P_{\max} = 0\,\text{dBW}$

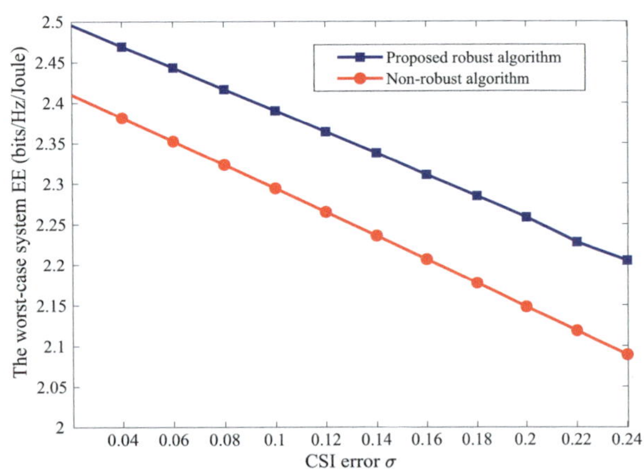

Fig. 6.6 The achievable worst-case system EE performance of the two designs as the functions of the CSI error threshold σ, given $N_t = 20$, $N_u = 2$, $K = 2$, $P_{\max} = 0\,\text{dBW}$, and $P_c = 0\,\text{dBW}$

6.6 Conclusion

In this chapter, a two timescale based dual-structured linear precoding scheme has been designed to realize the robust EE optimization in the dual-polarized multiuser MIMO system. The proposed algorithm can simultaneously reduce the CSI feedback overhead and the associated design operation complexity. Specifically, the system EE has been maximized by taking the norm-bounded CSI error into account in the dual-polarized MIMO system. Furthermore, the polarization-based subgrouping technique has been applied to all spatially grouped MUs to further reduce the CSI feedback overhead. Combining the two timescale based dual-structured precoding algorithm and the subgrouping algorithm, the optimal preprocessing matrix is first computed by exploiting the block diagonal structure of the spatial correlation matrix, and the linear precoding matrix is then optimized based on the instantaneous CSI via the proposed iterative algorithm. The performance advantages of the proposed algorithms are confirmed by simulation.

Appendix 1

Lemma 6.1 (sign-definiteness lemma) *For a Hermitian matrix* $\boldsymbol{Z}$ *and the arbitrary matrices* $\{\boldsymbol{X}_j, \boldsymbol{Y}_j, \boldsymbol{Q}_j\}_{j=1}^K$, *the semi-LMI with the following form*

$$\boldsymbol{Z} \succeq \sum_{j=1}^{K} (\boldsymbol{X}_j^{\mathrm{H}} \boldsymbol{Q}_j \boldsymbol{Y}_j + \boldsymbol{Y}_j^{\mathrm{H}} \boldsymbol{Q}_j^{\mathrm{H}} \boldsymbol{X}_j), \ \|\boldsymbol{Q}_j\| \leq \psi_j, \ \forall j, \tag{6.73}$$

holds if and only if there exists a set of nonnegative numbers $\xi_1, \ldots, \xi_K \in \mathbb{R}$ *satisfying*

$$\begin{bmatrix} \boldsymbol{Z} - \sum_{j=1}^{K} \xi_j \boldsymbol{X}_j^{\mathrm{H}} \boldsymbol{X}_j & -\psi_1 \boldsymbol{Y}_1^{\mathrm{H}} & \cdots & -\psi_K \boldsymbol{Y}_K^{\mathrm{H}} \\ -\psi_1 \boldsymbol{Y}_1 & \xi_1 \boldsymbol{I} & \cdots & \boldsymbol{0} \\ \vdots & \vdots & \ddots & \vdots \\ -\psi_K \boldsymbol{Y}_K & \boldsymbol{0} & \cdots & \xi_K \boldsymbol{I} \end{bmatrix} \succeq \boldsymbol{0}, \tag{6.74}$$

where all the matrices involved, including the identity matrix $\boldsymbol{I}$, *have appropriate dimensions.*

Appendix 2

Proof Firstly, we prove the convergence of the inner iterative procedure of Algorithm 6.1. For a given η, the objective function of the optimization problem (6.52) is given as follows,

$$\mathcal{J}\left(\boldsymbol{P}_k^p, f_{k,i}^p, \alpha_{k,i}^p\right) = \max_{\boldsymbol{\Delta}_{\mathcal{K}}} \sum_{k,i,p} \exp\left(\alpha_{k,i}^p - 1\right) MSE_{k,i}^p - \alpha_{k,i}^p + \eta \sum_{k,i,p} \left\| \boldsymbol{p}_k^{p,i} \right\|^2 + \eta P_c, \tag{6.75}$$

which is in fact equivalent to the optimization objective γ in the optimization problems (6.60) and (6.68). Let the solution obtained in the tth inner iteration of Algorithm 6.1 be $\left\{\boldsymbol{P}_k^{p(t)}, f_{k,i}^{p(t)}, \alpha_{k,i}^{p(t)}\right\}$. At the $(t+1)$th inner iteration, we first obtain the optimal $\alpha_{k,i}^{p(t+1)}$ based on $\boldsymbol{P}_k^{p(t)}$ and $f_{k,i}^{p(t)}$, which clearly satisfies

$$\mathcal{J}\left(\boldsymbol{P}_k^{p(t)}, f_{k,i}^{p(t)}, \alpha_{k,i}^{p(t)}\right) \geq \mathcal{J}\left(\boldsymbol{P}_k^{p(t)}, f_{k,i}^{p(t)}, \alpha_{k,i}^{p(t+1)}\right), \tag{6.76}$$

because the optimization (6.55) is a standard convex problem. Given $f_{k,i}^{p(t)}$ and $\alpha_{k,i}^{p(t+1)}$, the $\boldsymbol{P}_k^{p(t)}$ related optimization problem (6.60) is a standard SDP problem which has the globally optimal solution. Therefore, $\boldsymbol{P}_k^{p(t+1)}$ obtained in the $(t+1)$th iteration satisfies

$$\mathcal{J}\left(\boldsymbol{P}_k^{p(t)}, f_{k,i}^{p(t)}, \alpha_{k,i}^{p(t+1)}\right) \geq \mathcal{J}\left(\boldsymbol{P}_k^{p(t+1)}, f_{k,i}^{p(t)}, \alpha_{k,i}^{p(t+1)}\right). \tag{6.77}$$

Similarly, $f_{k,i}^{p(t+1)}$ obtained in the $(t+1)$th iteration, which is the solution of the SDP (6.68), satisfies

$$\mathcal{J}\left(\boldsymbol{P}_k^{p(t+1)}, f_{k,i}^{p(t)}, \alpha_{k,i}^{p(t+1)}\right) \geq \mathcal{J}\left(\boldsymbol{P}_k^{p(t+1)}, f_{k,i}^{p(t+1)}, \alpha_{k,i}^{p(t+1)}\right). \tag{6.78}$$

Combining (6.76) to (6.78), we have

$$\begin{aligned} \mathcal{J}\left(\boldsymbol{P}_k^{p(0)}, f_{k,i}^{p(0)}, \alpha_{k,i}^{p(0)}\right) &\geq \mathcal{J}\left(\boldsymbol{P}_k^{p(1)}, f_{k,i}^{p(1)}, \alpha_{k,i}^{p(1)}\right) \\ &\geq \cdots \geq \mathcal{J}\left(\boldsymbol{P}_k^{p(t)}, f_{k,i}^{p(t)}, \alpha_{k,i}^{p(t)}\right), \end{aligned} \tag{6.79}$$

which indicates that the objective function $\mathcal{J}\left(\boldsymbol{P}_k^p, f_{k,i}^p, \alpha_{k,i}^p\right)$ is non-increasing as t increases. Next, we prove that the objective function (6.52) becomes invariant with the increase of iterations. Note that the objective function (6.52) is equivalent to the negative objective of (6.48). For notational simplicity, we express (6.48) as

$$\widetilde{G}(\eta) = -\min_{f_{k,i}^p} \max_{\boldsymbol{\Delta}_{\mathcal{K}}} \sum_{k,i,p} \log\left(MSE_{k,i}^p\right) + \eta \sum_{k,i,p} \left\| \boldsymbol{p}_k^{p,i} \right\|^2 + \eta P_c = -\min_{f_{k,i}^p} \widetilde{\mathcal{G}}(\eta). \tag{6.80}$$

Then based on (6.50) and (6.51), the following equality holds

$$\min_{\alpha_{k,i}^k} \mathcal{J}\left(\boldsymbol{P}_k^p, f_{k,i}^p, \alpha_{k,i}^p\right) = \widetilde{\mathcal{G}}(\eta). \tag{6.81}$$

As aforementioned, $\widetilde{G}(\eta)$ is a lower bound of the objective $G(\eta)$ in the optimization problem (6.38), which is equivalent to the original worst-case EE optimization problem (6.34). Let us express $G(\eta)$ of (6.43) as

$$G(\eta) = -\max_{\mathbf{\Delta}_{\mathcal{K}}} \sum_{i,k,p} \log\left(\frac{1}{1+\mathrm{SINR}_{k,i}^{p}}\right) + \eta \sum_{i,k,p} \left\|\boldsymbol{p}_k^{p,i}\right\|^2 + \eta P_c = -\mathcal{G}(\eta). \quad (6.82)$$

Then, for any optimal $\boldsymbol{P}_k^p$, $f_{k,i}^p$, $k \in \mathcal{K}$, $p \in \{h, v\}$, we have

$$G(\eta) = -\mathcal{G}(\eta) \geq \widetilde{G}(\eta) = -\widetilde{\mathcal{G}}(\eta), \quad (6.83)$$

that is,

$$\mathcal{G}(\eta) \leq \widetilde{\mathcal{G}}(\eta). \quad (6.84)$$

Noting (6.81) and (6.84), for the given η, we have

$$\mathcal{J}\left(\boldsymbol{P}_k^p, f_{k,i}^p, \alpha_{k,i}^p\right) \geq \min_{\alpha_{k,i}^p} \mathcal{J}\left(\boldsymbol{P}_k^p, f_{k,i}^p, \alpha_{k,i}^p\right) \geq \mathcal{G}(\eta). \quad (6.85)$$

We now consider to derive a lower bound of $\mathcal{G}(\eta)$. Recall the SINR definition of (6.29). It is clear that an upper bound of this $SINR_{k,i}^p$ is given by

$$\mathrm{SINR}_{k,i}^p \leq \left|\boldsymbol{h}_{kk,i}^p \boldsymbol{p}_k^{p,i}\right|^2 \leq \left\|\boldsymbol{h}_{kk,i}^p\right\|^2 \left\|\boldsymbol{p}_k^{p,i}\right\|^2 \leq \left(\left\|\widehat{\boldsymbol{h}}_{kk,i}^p\right\| + \tau_{kk,i}^p\right)^2 P_{\max}, \quad (6.86)$$

where the last inequality holds due to the channel error model (6.32) and (6.33) as well as the fact that $\left\|\boldsymbol{p}_k^{p,i}\right\|^2 \leq \|\boldsymbol{P}\|_F^2 \leq P_{\max}$. Substituting $SINR_{k,i}^p$ with its upper bound (6.86) in $\mathcal{G}(\eta)$ leads to the following lower bound of $\mathcal{G}(\eta)$

$$\begin{aligned}\mathcal{G}(\eta) &= \max_{\mathbf{\Delta}} \sum_{i,k,p} \log\left(\frac{1}{1+\mathrm{SINR}_{k,i}^p}\right) + \eta \sum_{i,k,p} \left\|\boldsymbol{p}_k^{p,i}\right\|^2 + \eta P_c \\ &\geq \max_{\mathbf{\Delta}_{\mathcal{K}}} \sum_{i,k,p} \log\left(\frac{1}{1+\mathrm{SINR}_{k,i}^p}\right) \\ &\geq \sum_{i,k,p} \log\left(\frac{1}{1+\left(\left\|\widehat{\boldsymbol{h}}_{kk,i}^p\right\| + \tau_{kk,i}^p\right)^2 P_{\max}}\right).\end{aligned} \quad (6.87)$$

Moreover, according to (6.85), (6.87) equivalently acts as a lower bound to $\mathcal{J}\left(\boldsymbol{P}_k^p, f_{k,i}^p, \alpha_{k,i}^p\right)$. Combining the non-increasing and lower bounded characteristics of the objective function $\mathcal{J}\left(\boldsymbol{P}_k^p, f_{k,i}^p, \alpha_{k,i}^p\right)$, we conclude that the inner iteration procedure of Algorithm 6.1 converges.

The outer iteration procedure of updating η is performed by the bisection method, which is guaranteed to converge. Therefore, the convergence of Algorithm 6.1 is guaranteed.

Lastly, we prove that Algorithm 6.1 converges to a local optimal solution of the optimization problem (6.52). Assume that Algorithm 6.1 converges to the solution

$$\boldsymbol{V}^{\star} = \left\{ \boldsymbol{P}_k^{p\star}, f_{k,i}^{p\star}, \alpha_{k,i}^{p\star} \right\}. \tag{6.88}$$

Now consider an arbitrary point $\boldsymbol{V} = \left\{ \boldsymbol{P}_k^{p}, f_{k,i}^{p}, \alpha_{k,i}^{p} \right\}$ in the neighbourhood of $\boldsymbol{V}^{\star}$, that is,

$$\boldsymbol{V} \in \mathcal{V} = \left\{ \tilde{\boldsymbol{V}} : \left\| \tilde{\boldsymbol{V}} - \boldsymbol{V}^{\star} \right\|_F \leq \zeta \right\}, \tag{6.89}$$

where $\zeta > 0$ is a sufficiently small positive number. In the proposed iterative algorithm for solving the optimization problem (6.52), each of the three optimization subproblems is a standard convex optimization with the globally optimal solution. Therefore, for any $\boldsymbol{V} \in \mathcal{V}$, we have

$$\operatorname{Tr}\left(\nabla_{\boldsymbol{P}_k^{p\star}} \mathcal{J}\left(\boldsymbol{P}_k^{p}, f_{k,i}^{p\star}, \alpha_{k,i}^{p\star} \right) \left(\boldsymbol{P}_k^{p} - \boldsymbol{P}_k^{p\star} \right) \right) \geq 0, \tag{6.90}$$

$$\operatorname{Tr}\left(\nabla_{f_{k,i}^{p\star}} \mathcal{J}\left(\boldsymbol{P}_k^{p\star}, f_{k,i}^{p}, \alpha_{k,i}^{p\star} \right) \left(f_{k,i}^{p} - f_{k,i}^{p\star} \right) \right) \geq 0, \tag{6.91}$$

$$\operatorname{Tr}\left(\nabla_{\alpha_{k,i}^{p\star}} \mathcal{J}\left(\boldsymbol{P}_k^{p\star}, f_{k,i}^{p\star}, \alpha_{k,i}^{p} \right) \left(\alpha_{k,i}^{p} - \alpha_{k,i}^{p\star} \right) \right) \geq 0, \tag{6.92}$$

where $\nabla_A \mathcal{J}$ denotes the gradient of $\mathcal{J}$ with respect to A. By adding up all the above three inequalities, we have

$$\operatorname{Tr}\left(\nabla_{\boldsymbol{V}^{\star}} \mathcal{J}\left(\boldsymbol{V}^{\star} \right) \left(\boldsymbol{V} - \boldsymbol{V}^{\star} \right) \right) \geq 0. \tag{6.93}$$

Because $\mathcal{J}$ is a continuous function, we conclude

$$\mathcal{J}(\boldsymbol{V}) - \mathcal{J}\left(\boldsymbol{V}^{\star} \right) \geq 0. \tag{6.94}$$

This proves that $\boldsymbol{V}^{\star} = \left\{ \boldsymbol{P}_k^{p\star}, f_{k,i}^{p\star}, \alpha_{k,i}^{p\star} \right\}$ is a stationary point.

References

1. Kim T, Clerckx B, Love DJ et al (2010) Limited feedback beamforming systems for dual-polarized MIMO channels. IEEE Trans Wireless Commun 9(11):3425–3439
2. Ben Zid M, Raoof K, Bouallegue A (2012) Dual polarized versus single polarized MIMO: a study over NLOS propagation with polarization discrimination and spatial correlation effects. In: European conference on antennas and propagation. IEEE, pp 1979–1983

3. Gong S, Xing C, Chen S et al (2017) Secure communications for dual-polarized MIMO systems. IEEE Trans Signal Process 65(16):4177–4192
4. Dong L, Choo H, Heath RW et al (2005) Simulation of MIMO channel capacity with antenna polarization diversity. IEEE Trans Wireless Commun 4(4):1869–1873
5. Yang Z, Warnick KF (2014) Effect of mutual coupling on the sensitivity of dual polarized receivers in satellite communications. In: IEEE antennas and propagation society international symposium. IEEE, pp 1528–1529
6. Gong S, Xing C, Chen S et al (2018) Polarization sensitive array based physical-layer security. IEEE Trans Veh Technol 67(5):3964–3981
7. Oestges C, Erceg V, Paulraj AJ (2004) Propagation modeling of MIMO multi-polarized fixed wireless channels. IEEE Trans Veh Technol 53(3):644–654
8. Oestges C, Clerckx B, Guillaud M et al (2008) Dual-polarized wireless communications: from propagation models to system performance evaluation. IEEE Trans Wireless Commun 7(10):4019–4031
9. Nabar RU, Bolcskei H, Erceg V et al (2002) Performance of multiantenna signaling techniques in the presence of polarization diversity. IEEE Trans Signal Process 50(10):2553–2562
10. Liu Y, Guo C, Zeng Z, et al (2014) The projection-based self-interference cancellation scheme in the polarization domain. In: International symposium on wireless personal multimedia communications. IEEE, pp 408–413
11. Kim T, Clerckx B, Love DJ et al (2008) Limited feedback beamforming codebook design for dual-polarized MIMO channels. In: IEEE Global Telecommunications Conference. IEEE, pp 1–5
12. Park J, Clerckx B (2015) Multi-user linear precoding for multi-polarized massive MIMO system under imperfect CSIT. IEEE Trans Wireless Commun 14(5):2532–2547
13. He S, Huang Y, Yang L et al (2015) Energy efficient coordinated beamforming for multicell system: Duality-based algorithm design and massive MIMO transition. IEEE Trans Commun 63(12):4920–4935
14. He S, Huang Y, Wang H et al (2014) Leakage-aware energy-efficient beamforming for heterogeneous multicell multiuser systems. IEEE J Sel Areas Commun 32(6):1268–1281
15. Li Y, Fan P, Beaulieu NC (2016) Cooperative downlink max-min energy efficient precoding for multicell MIMO networks. IEEE Trans Veh Technol 65(11):9425–9430
16. Zhang Z, Liu W, Leng W et al (2016) Interference-plus-Noise covariance matrix reconstruction via spatial power spectrum sampling for robust adaptive beamforming. IEEE Signal Process Lett 23(1):121–125
17. Huang L, Zhang J, Xu X et al (2015) Robust adaptive beamforming with a novel interference-plus-noise covariance matrix reconstruction method. IEEE Trans Signal Process 63(7):1643–1650
18. Ruan H, de Lamare RC (2016) Robust adaptive beamforming based on low-rank and cross-correlation techniques. IEEE Trans Signal Process 64(15):3919–3932
19. Yang L, McKay MR, Couillet R (2018) High-dimensional MVDR beamforming: Optimized solutions based on spiked random matrix models. IEEE Trans Signal Process 66(7):1933–1947
20. Chakrabarty S, Habets EAP (2018) A Bayesian approach to informed spatial filtering with robustness against DoA estimation errors. IEEE/ACM Trans Audio Speech Lang Process 26(1):145–160
21. Zappone A, Cao P, Jorswieck EA (2014) Low-complexity energy efficiency optimization with statistical CSI in two-hop MIMO systems. IEEE Signal Process Lett 21(11):1398–1402
22. Zhang C, Ge J, Li J et al (2016) Energy efficiency and spectral efficiency tradeoff for asymmetric two-way AF relaying with statistical CSI. IEEE Trans Veh Technol 65(4):2833–2839
23. Wang L, Sheng M, Zhang Y et al (2015) Robust energy efficiency maximization in cognitive radio networks: The worst-case optimization approach. IEEE Trans Commun 63(1):51–65
24. Cui Y, Xu W, Shen H et al (2016) CSI impaired precoding optimization for energy-efficient MIMO communications under total power constraint. IEEE Commun Lett 20(3):514–517
25. Mao Y, Clerckx B, Li VOK (2018) Energy efficiency of rate-splitting multiple access, and performance benefits over SDMA and NOMA. In: International symposium on wireless communication systems. IEEE, pp 1–5

26. Dai M, Clerckx B, Gesbert D et al (2016) A rate splitting strategy for massive MIMO with imperfect CSIT. IEEE Trans Wireless Commun 15(7):4611–4624
27. Joudeh H, Clerckx B (2016) Robust transmission in downlink multiuser MISO systems: a rate-splitting approach. IEEE Trans Signal Process 64(23):6227–6242
28. Chen J, Pratt T (2014) Energy efficiency of adaptive transmission for MIMO-OFDM systems over polarization-sensitive channels. In: National wireless research collaboration symposium. IEEE, pp 127–132
29. Chen J, Pratt TG (2014) Energy efficiency of space and polarization MIMO communications with packet erasures over wireless fading channels. IEEE Trans Wireless Commun 13(12):6557–6569
30. Coldrey M (2008) Modeling and capacity of polarized MIMO channels. In: IEEE Vehicular Technology Conference Spring. IEEE, pp 440–444
31. Gong S, Xing C, Yang N et al (2017) Energy efficient transmission in multi-user MIMO relay channels with perfect and imperfect channel state information. IEEE Trans Wireless Commun 16(6):3885–38989
32. Kumar RVR, Gurugubelli J (2011) How green the LTE technology can be? In: International conference on wireless communication, vehicular technology, information theory and aerospace and electronic systems technology. IEEE, pp 1–5
33. Adhikary A, Nam J, Ahn J-Y et al (2013) Joint spatial division and multiplexing-The large-scale array regime. IEEE Trans Inf Theory 59(10):6441–6463
34. Aubry A, Carotenuto V, Maio AD (2016) New results on generalized fractional programming problems with Toeplitz quadratics. IEEE Signal Process Lett 23(6):848–852
35. Dinkelbach W (1967) On nonlinear fractional programming. Manag Sci 13(7):492–498
36. Shi Q, Razaviyayn M, Luo Z-Q et al (2011) An iteratively weighted MMSE approach to distributed sum-utility maximization for a MIMO interfering broadcast channel. IEEE Trans Signal Process 59(9):4331–4340
37. Christensen SS, Argawal R, de Carvalho E et al (2008) Weighted sum-rate maximization using weighted MMSE for MIMO-BC beamforming design. IEEE Trans Wireless Commun 7(12):4792–4799
38. Xu W, Cui Y, Zhang H et al (2015) Robust beamforming with partial channel state information for energy efficient networks. IEEE J Sel Areas Commun 33(12):2920–2935
39. Ben-Tal A, Nemirovski A (2001) Lectures on modern convex optimization: analysis, algorithms, and engineering applications. SIAM
40. Gharavol EA, Larsson EG (2013) The sign-definiteness lemma and its applications to robust transceiver optimization for multiuser MIMO systems. IEEE Trans Signal Process 61(2):238–252
41. Boyd S, Vandenberghe L (2004) Convex optimization. Cambridge University Press
42. Shiu D-S, Foschini GJ, Gans MJ et al (1998) Fading correlation and its effect on the capacity of multi-element antenna systems. In: IEEE International Conference on Universal Personal Communications. IEEE, pp 429–433

Chapter 7
Energy-Efficient Beamforming for Satellite MIMO Transmission

Jiaxuan Li, Xuhui Ding, and Neng Ye

In this chapter, we study the EE optimization problem of MIMO multi-satellite uplink communication systems based on lens antenna arrays. Section 7.1 introduces the research status of EE in multi-satellite uplink transmission. The system model and the problem formulation are introduced in Sect. 7.2. Section 7.3 provides the algorithm that jointly optimizes analog beam selection and baseband digital precoding. Section 7.4 introduces the iteration optimization algorithm that optimizes the number of required RF chains. In Sect. 7.5, the system performance is analyzed. Finally, conclusions are drawn in Sect. 7.6.

7.1 Introduction

The lens antenna array is a promising solution for multi-satellite uplink transmission [1, 2]. Different from other types of multi-beam antennas that have already been applied in satellite-to-ground communication, including reflector antenna and phased array antenna [3, 4], lens antenna array allows for wide-angle fast beam-scanning and low-sidelobe beamforming at a relatively low cost with low energy consumption [5]. These properties make it capable of meeting the stringent power constraint on ground terminals, and adapting to satellites with various tilt angles in dense Earth orbit constellations. Due to its superiority in satellite communication, Luneburg lens

J. Li · X. Ding · N. Ye (✉)
Beijing Institute of Technology, Beijing, China
e-mail: ianye@bit.edu.cn

J. Li
e-mail: lijiaxuan@bit.edu.cn

X. Ding
e-mail: dingxuhui@bit.edu.cn

X. Ding et al. (eds.), *Performance Analysis and Improvement in MIMO Communication Systems*, https://doi.org/10.1007/978-981-96-5951-7_7

antenna array has been widely used in satellite information gathering, ship-satellite communication systems, as well as mobile satellite communication [6–8], with an inter-beam isolation[1] of more than 30 dB and high gain multi-beam coverage of a 120° field-of-view [9, 10].

Although lens antenna array has the advantage of low power consumption, EE still remains a challenge in multi-satellite uplink transmission. In such systems, the ground terminals are sometimes on the sea, in the desert or in other unmanned areas, far away from the power plant. As a result, the terminals are subject to energy constraints that require them to operate on limited power. Moreover, in the scenarios that we consider, according to the fast development of Earth orbit satellites, it can be indicated that the terminal may connect to more satellites over long propagation distance, which means that the effect of path loss tends to be greater, leading to much more energy consumption. Considering the multi-satellite system we focus on, the ground terminals equipped with lens antennas should connect to multiple satellites simultaneously, where the energy consumption is decided by the transmission parameters. Unreasonable parameter configurations such as the inappropriate number of RF chains and incorrectly setting of RF chain selection matrix will lead to extra energy consumed. According to the facts above, it is necessary for the ground terminals to adjust the transmission parameters to comprehensively consider the process of precoding and RF chain selection, and achieve a balance between total data rate and EE, so as to increase the operation time under constrained energy supply.

Taking EE as the major performance metric, many studies have investigated the energy-efficient transmission strategy in various wireless communication systems [11, 12], while the introduction of lens antenna is hardly considered, despite its benefits on energy consumption and wide angular coverage. However, as the available frequency spectrum is also limited [13], most of the studies on wireless communication systems with lens antenna arrays focus on optimizing SE or sum rate [14–17] rather than EE, which leads to the inefficiency of these schemes for energy-constrained transmission systems.

In the research on optimizing EE in a lens antenna array based system, joint design of baseband digital precoding and the number of required RF chains is the lack of consideration. Authors in [18] propose an IRSs-enhanced mmWave lens system, where the lens antenna is used to reduce the number of required RF chains during the alternating optimization of reflective and transmit beamforming. Aiming to break the fundamental limit on the number of served users to be less than the number of RF chains in existing beamspace MIMO systems, authors in [19] propose a beamspace MIMO non-orthogonal multiple access (NOMA) mmWave system, where no optimization is performed on the number of required RF chains. Authors in [20] propose a sum-rate and EE optimization scheme for a wideband mmWave communication system, in which the problem is decoupled into a beamspace baseband digital precoding optimization problem and analog beam selection optimization problem. As designing the analog beam selection and baseband digital precoding independently

[1] Inter-beam isolation characterizes the magnitude of the crosstalk between any two beams, and larger inter-beam isolation provides better signal quality.

is unlikely to be optimal, there still needs to be a joint design of these processes. Machine learning is a feasible solution for a joint design of analog beam selection and baseband digital precoding. Authors in [21, 22] take cross-entropy in machine learning into consideration and propose hybrid precoding schemes to achieve higher sum-rate and EE. However, although there has been researches that consider the EE optimization in a beamspace system, to the best of our knowledge, none of the reviewed works on lens antenna array have optimized the number of required RF chains under the usage of lens antenna array in multi-satellite transmission scenarios for EE optimization.

Notation: Normal lower-case, bold lower-case and bold upper-case symbols denote scalars, vectors and matrices, respectively. $\mathbb{R}$ and $\mathbb{C}$ denote the fields of real and complex numbers, respectively. $|\cdot|$ denotes the modulus of a scalar or the cardinality of a set, and $\|\cdot\|$ represents the F norm. $\mathcal{CN}(\boldsymbol{\mu}, \boldsymbol{\Sigma})$ represents circular symmetric complex Gaussian distribution with mean $\boldsymbol{\mu}$ and covariance $\boldsymbol{\Sigma}$.

7.2 System Model

As shown in Fig. 7.1, we consider a multi-satellite uplink communication system based on lens antennas, which includes a ground station and several single-antenna satellites. The ground station has a multi-antenna structure which uses the lens antenna array. The signals are transmitted by N_{RF} RF chains on the surface of the lens antenna array, and the spherical wave on the lens surface is collimated into a plane beam in a specific direction. Since any point on the surface of the lens can be regarded as the focus, the multi-satellite connections can be realized by placing multiple feeds on the lens antenna.

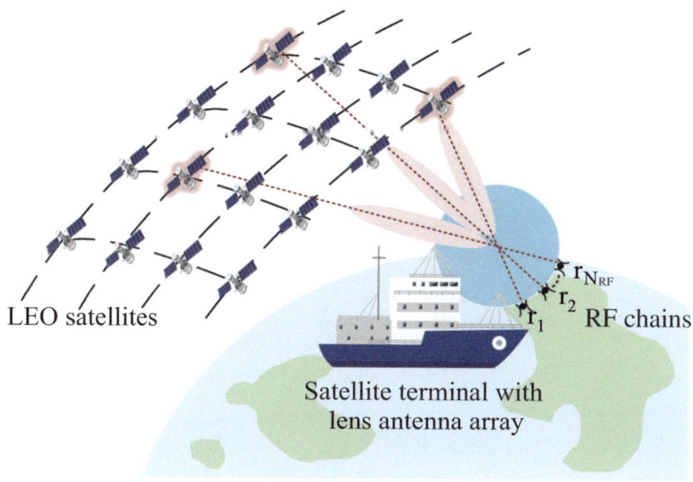

Fig. 7.1 The general framework of lens antenna array based multi-satellite uplink system

In the following parts, we first construct the multi-satellite transmission system model. Then, the inside structure and the parameters of lens antennas are introduced. Based on the lens antenna array equations, the channel model is further expressed. Finally, the EE optimization problem is formulated.

7.2.1 *Lens Antenna Array Based Multi-satellite Uplink System Model*

Generally, as shown in Fig. 7.2, there are N_{T} elements mounted on the lens antenna array and N_{RF} RF chains at the ground station, which processes and transmits K independent data streams to the corresponding K satellites. We assume $K \leq N_{\mathrm{RF}} \leq N_{\mathrm{T}}$. We consider an orthogonal multiple access among beams and satellites.

7.2.1.1 Channel Model

There is one LoS path component and L NLoS path components in the channel coefficient between the ground station and the kth satellite. We define a satellite space direction as the set of all path directions related to this satellite, considering a uniform linear array for antennas on the lens surface, we have

$$\mathbf{u}(\theta_{k,l}) = \frac{1}{\sqrt{N_{\mathrm{T}}}}\left[1, e^{j2\pi\theta_{k,l}}, \ldots, e^{j(N_{\mathrm{T}}-1)2\pi\theta_{k,l}}\right]^T, \tag{7.1}$$

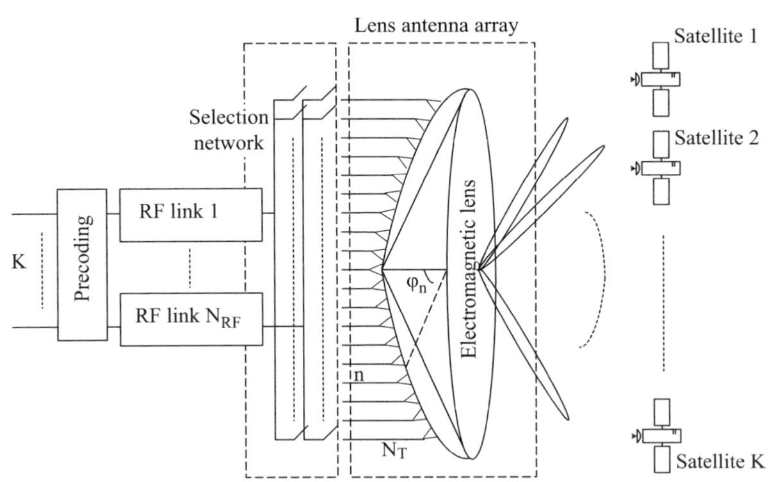

Fig. 7.2 System framework of lens antenna array based multi-satellite uplink system

where $\theta_{k,l}$ represents the spatial direction corresponding to the lth path of the kth satellite. It can be expressed as $\theta_{k,l} \triangleq \frac{d}{\lambda}\sin(\phi_{k,l})$, where $\phi_{k,l}$ is the physical direction angle between the ground terminal and the lth path of the kth satellite. λ is the signal wavelength and d is the interval between adjacent array elements of the antenna array. We set $d = \frac{\lambda}{2}$ following the general setting [23]. j is the imaginary unit.

For the mmWave communication channel, we consider the Saleh-Valenzuela model [24, 25], in which $\beta_{k,0}$ and $\beta_{k,l}$ represent the gain of the LoS and NLoS path respectively. $\beta_{k,0}\mathbf{u}(\theta_{k,0})$ and $\beta_{k,l}\mathbf{u}(\theta_{k,l}), l = 1, \ldots, L$ respectively represent the corresponding path components. The channel vector between the ground station and the kth satellite $\mathbf{h}_k \in \mathbb{C}^{N_\mathrm{T}\times 1}$ can be expressed as

$$
\begin{aligned}
\mathbf{h}_k &= \sqrt{\frac{N_\mathrm{T}}{L+1}} \sum_{l=0}^{L} \beta_{k,l}\mathbf{u}(\theta_{k,l}) \\
&= \sqrt{\frac{N_\mathrm{T}}{L+1}}\beta_{k,0}\mathbf{u}(\theta_{k,0}) + \sqrt{\frac{N_\mathrm{T}}{L+1}} \sum_{l=1}^{L} \beta_{k,l}\mathbf{u}(\theta_{k,l}).
\end{aligned}
\tag{7.2}
$$

Therefore, $\mathbf{H} = [\mathbf{h}_1, \ldots, \mathbf{h}_k, \ldots, \mathbf{h}_K] \in \mathbb{C}^{K\times N_\mathrm{T}}$ indicates the uplink channel matrix of the entire communication system.

7.2.1.2 Signal Model

The lens antenna array is composed of an electromagnetic lens with energy-concentrating effect and a series of matched antennas. The electromagnetic lens samples the incident signal, and adds an appropriate delay. In this way, the signals from all directions can be converged on the focal arc of the lens. Through the design of the lens, the lens antenna array can realize the function of spatial DFT [26–28]. The receiving response of the nth antenna is

$$
\mathbf{a}(\varphi_n) = \frac{1}{\sqrt{N_\mathrm{T}}}[1, e^{j2\pi\varphi_n}, \ldots, e^{j(N_\mathrm{T}-1)2\pi\varphi_n}]^T. \tag{7.3}
$$

Assuming N_T is an odd number, $\varphi_n = \frac{1}{N_\mathrm{T}}(n - \frac{N_\mathrm{T}+1}{2}), n = 1, \ldots, N_\mathrm{T}$ represents a preset spatial direction relative to the nth antenna. $\frac{1}{N_\mathrm{T}}$ gives the spatial direction resolution. That is to say, the receiving array response $\mathbf{A} \in \mathbb{C}^{N_\mathrm{T}\times N_\mathrm{T}}$ is

$$
\mathbf{A} = \left[\mathbf{a}(\varphi_1), \ldots, \mathbf{a}(\varphi_{N_\mathrm{T}})\right]^H. \tag{7.4}
$$

Since the electromagnetic lens is a passive device, the reciprocity holds between its receive array response and transmit array response [29]. Therefore, the transmit array response can be given by $\mathbf{A}^H \in \mathbb{C}^{N_\mathrm{T}\times N_\mathrm{T}}$. It maps the signal on each antenna to the beam in its corresponding direction $\mathbf{a}(\varphi_n)$, which is the column vector of $\mathbf{A}^H$.

Meanwhile, considering the sparsity of the mmWave channel, we can select specific beams in the communication. In the mathematical expression, we can use the analog beam selection matrix $\mathbf{S}_{\mathrm{RF}} \in \mathbb{R}^{N_{\mathrm{T}} \times N_{\mathrm{RF}}}$ to select the corresponding column vector in the $\mathbf{A}^H$. We denote the nth column of $\mathbf{S}_{\mathrm{RF}}$ as $\mathbf{s}_{\mathrm{RF,n}}$. For example, if we choose the 3rd and 5th beams, there is

$$\mathbf{S}_{\mathrm{RF}} = [\mathbf{e}_3\ \mathbf{e}_5] \in \mathbb{R}^{N_{\mathrm{T}} \times 2}, \tag{7.5}$$

where $\mathbf{e}_i \in \mathbb{R}^{N_{\mathrm{T}} \times 1}$ is a column vector with the ith element being 1 and the other elements being 0.

Combining the lens antenna array response and channel model, the received signal of satellites $\mathbf{y} \in \mathbb{C}^{K \times 1}$ can be expressed as

$$\mathbf{y} = \mathbf{H}^H \mathbf{A}^H \mathbf{S}_{\mathrm{RF}} \mathbf{P} \mathbf{s} + \mathbf{n}, \tag{7.6}$$

where the kth element $y_k, k = 1, \ldots, K$ in $\mathbf{y}$ represents the received signal of the kth satellite. $\mathbf{P} \in \mathbb{C}^{N_{\mathrm{RF}} \times K}$ represents the baseband digital precoding matrix. $\mathbf{s} \in \mathbb{C}^{K \times 1}$ represents a vector of information sent to the satellite, and there is $\mathbf{s} \sim \mathcal{CN}(0, \mathbf{I}_K)$. We define the information after baseband digital precoding as $\mathbf{r} = \mathbf{P}\mathbf{s} \in \mathbb{C}^{N_{\mathrm{RF}} \times 1}$, where the ith element $r_i, i = 1, \ldots, N_{\mathrm{RF}}$ in $\mathbf{r}$ denotes the information on the ith RF chain in Fig. 7.1. $\mathbf{n} \in \mathbb{C}^{K \times 1}$ is the AWGN with variance σ_n^2, and there is $\mathbf{n} \sim \mathcal{CN}(0, \sigma_n^2 \mathbf{I}_K)$. Since analog beam selection corresponds to antenna selection while each antenna corresponds to an RF chain, the number of beams is equal to the number of required RF chains.

At the same time, the valid signal power sent to the kth satellite is

$$P_k = \left\| \mathbf{A}^H \mathbf{S}_{\mathrm{RF}} \mathbf{p}_k s_k \right\|^2 = \|\mathbf{p}_k\|^2, \tag{7.7}$$

where $\mathbf{p}_k \in \mathbb{C}^{N_{\mathrm{RF}} \times 1}$ is the kth column of $\mathbf{P}$. The equation holds because $\mathbf{A}^H \mathbf{A} = \mathbf{A}\mathbf{A}^H = \mathbf{I}_{N_{\mathrm{T}}}$ and $\mathbf{S}_{\mathrm{RF}}^H \mathbf{S}_{\mathrm{RF}} = \mathbf{I}_{N_{\mathrm{RF}}}$.

Combined with the lens antenna array, the response of the different beams of the lens antenna array to the kth satellite space direction can be expressed as

$$\tilde{\mathbf{h}}_k = \mathbf{A}\mathbf{h}_k. \tag{7.8}$$

Therefore, the channel matrix is transformed into a beam space channel matrix [26]. $\tilde{\mathbf{H}} = \left[\tilde{\mathbf{h}}_1, \ldots, \tilde{\mathbf{h}}_k, \ldots, \tilde{\mathbf{h}}_K\right]$. It provides a more insightful way to describe the response amplitude of the antennas of the ground terminal. The data rate of the kth satellite is

$$R_k = \log_2 (1 + \gamma_k), \tag{7.9}$$

where

$$\gamma_k = \frac{\left| \tilde{\mathbf{h}}_k^H \mathbf{S}_{\mathrm{RF}} \mathbf{p}_k \right|^2}{\sum\limits_{i=1, i \neq k}^{K} \left| \tilde{\mathbf{h}}_k^H \mathbf{S}_{\mathrm{RF}} \mathbf{p}_i \right|^2 + \sigma_n^2}. \tag{7.10}$$

Suggesting that we only consider LoS path components, $\mathbf{h}_k^{'}$ can be expressed from (7.2)

$$\mathbf{h}_k^{'} = \sqrt{N_\mathrm{T}}\beta_{k,0}\mathbf{u}(\theta_{k,0}) = \sqrt{N_\mathrm{T}}\beta_k\mathbf{u}(\theta_k). \tag{7.11}$$

Then the nth component of $\tilde{\mathbf{h}}_k^{'}$ can be obtained according to (7.4) as

$$\tilde{\mathbf{h}}_{k,n}^{'} = \beta_k e^{j\pi(N_\mathrm{T}-1)(\theta_k-\varphi_n)} \frac{\sin N\pi(\theta_k - \varphi_n)}{\sqrt{N}\sin\pi(\theta_k - \varphi_n)}. \tag{7.12}$$

Figure 7.3 gives the normalized beamspace channel response magnitudes, i.e., $\left|\tilde{\mathbf{h}}_{k,n}\right|$, for different spatial directions with $N_\mathrm{T} = 13$. Equation (7.12) and Fig. 7.3 show that when the satellite space direction is consistent with the preset space direction of the lens antenna array, the signal on this antenna will obtain the maximum gain. The signal gain is 0 on other antennas in this direction due to the orthogonality of the DFT matrix. However, in actual transmission systems, it is unrealistic to match all satellite space directions. Therefore, the main beams that the gain of a direction concentrates on can be selected for transmission. However, the interference between satellites can be relatively serious when the direction difference is within $1/(2N_\mathrm{T})$. This problem can be overcome by increasing the number of antennas to improve the spatial direction resolution or adding additional interference cancellation algorithms.

In addition, due to the short wavelength of mmWave [30], mmWave signals are prone to attenuation and are sensitive to obstruction and scattering, so the signal propagation path is limited in mmWave communication systems and the beamspace channel matrix $\tilde{\mathbf{H}}$ is sparse. Thus, we can select the main beams for communication, which decreases the number of required RF chains and reduces the complexity of system implementation. It is worth mentioning that although the short wavelength and the long propagation distance bring large path loss for the usage of mmWave in low Earth orbit satellite communications, the severe attenuation can be covered by employing high-gain directional antennas to achieve high-capacity satellite communication, which shows the feasibility of using mmWave in satellite communications [31].

Fig. 7.3 Normalized beamspace channel response amplitude

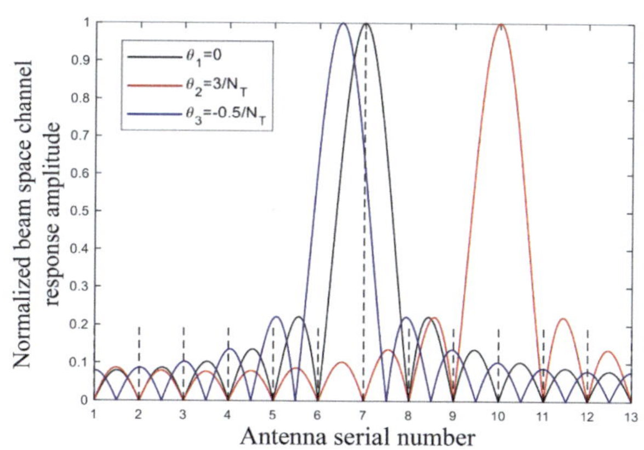

7.2.2 Problem Formulation

7.2.2.1 Power Consumption Model

To formulate the EE optimization problem, we consider the components of the beamforming structure. As we know, the analog beam selection network in Fig. 7.2 is composed of analog switches and each RF chain needs to connect to specific antennas from the total N_T antennas. Hence, the total system power consumption can be expressed as

$$P_{\text{total}} = \frac{1}{\alpha}\sum_{k=1}^{K} P_k + N_{\text{RF}}(P_{\text{RF}} + N_{\text{T}} P_{\text{SW}}) + P_{\text{C}}, \tag{7.13}$$

where α represents the power amplifier efficiency, P_{RF} and P_{SW} are the RF chain power consumption and the analog switch power consumption, respectively. The power consumption of the analog switch is less than the power consumption of the traditional phase shifter [32]. P_{C} represents the power consumption of other hardware components in the system.

Based on the above discussion, we can give the system EE expression as

$$EE = \frac{\sum_{k=1}^{K} R_k}{P_{\text{total}}}. \tag{7.14}$$

7.2.2.2 Problem Formulation

According to the previous discussion, in order to maximize the EE of the system, we need to design the baseband digital precoding matrix **P**. Meanwhile, the analog beam selection matrix and the number of required RF chains are optimized according to the beamspace channel information. Therefore, the EE optimization problem of a multi-satellite communication system based on a lens antenna array is formulated as follows.

$$\mathcal{P}1 : \left(N_{\text{RF}}^{opt}, \mathbf{S}_{\text{RF}}^{opt}, \mathbf{P}^{opt}\right) = \arg\max_{N_{\text{RF}}, \mathbf{S}_{\text{RF}}, \mathbf{P}} \; EE \tag{7.15}$$

$$\text{s.t.} \quad \sum_{k=1}^{K} P_k \leqslant P_{\max}, \tag{7.15a}$$

$$P_k \geqslant 0, \tag{7.15b}$$

$$\gamma_k \geqslant \gamma_{\min}, \tag{7.15c}$$

$$\mathbf{s}_{\text{RF,m}} \in \{\mathbf{e}_i, i = 1, \ldots, N_{\text{T}}\},$$

$$\mathbf{s}_{\mathrm{RF,m}} \neq \mathbf{s}_{\mathrm{RF,n}}, \forall m \neq n, m, n = 1, \ldots, N_{\mathrm{RF}}. \tag{7.15d}$$

In (7.15), the constraints (7.15a) and (7.15b) represent the uplink transmit power budget constraints for the satellites, where $P_{\max}$ is the maximum signal transmit power. Besides, the constraint (7.15c) represents the satellites' requirement for the received SINR. Since the data rate per unit bandwidth R_k, the spectral efficiency, is a monotonically increasing function of SINR, the constraint on SINR is equivalent to the constraint on spectral efficiency. Finally, the constraint (7.15d) stands for the analog beam selection constraint, which implies that each antenna can only be connected to one RF chain, and each RF chain can only be connected to one antenna.

The problem ($\mathcal{P}1$) is maximizing a non-concave objective function and its constraints are non-convex. Based on the definitions of the analog beam selection matrix $\mathbf{S}_{\mathrm{RF}}$ and the baseband digital precoding matrix $\mathbf{P}$, the dimensions of these two matrices are related to the number of required RF chains N_{RF}. So it is difficult to obtain the global optimal solution of the problem. To solve this problem, we propose a feasible suboptimal method. In order to simplify the derivation, we first jointly optimize the analog beam selection and baseband digital precoding given a number of required RF chains. Then, based on the proposed method for analog beam selection and baseband digital precoding, the optimal number of required RF chains is calculated to maximize the system EE. The algorithm for searching the optimal number of RF chains is a global optimization algorithm, so the choice of the given number of RF chains does not influence the final results.

7.3 Analog Beam Selection and Digital Baseband Precoding Design

In this section, we first discuss the problem transformation based on a hybrid analog-digital precoding method. Then, we obtain the fully-digital precoding matrix. Finally, the Euclidean distance is considered to make the hybrid analog-digital precoding closer to the fully digital one, thus completing the optimization design of an analog beam selection matrix and the digital baseband precoding matrix.

7.3.1 Problem Transformation Based on Hybrid Analog-Digital Precoding Method

Based on what has been stated hitherto, we first assume a given number of RF chains. Then, the EE optimization problem can be expressed as follows.

$$\mathcal{P}2 : \left(\mathbf{S}_{\mathrm{RF}}^{opt}, \mathbf{P}^{opt}\right) = \arg\max_{\mathbf{S}_{\mathrm{RF}}, \mathbf{P}} \quad EE \tag{7.16}$$

$$\text{s.t.} \quad (7.15a), (7.15b), (7.15c), (7.15d).$$

But the problem ($\mathcal{P}2$) is still difficult to solve. We consider the form of lens antenna array response matrix and combine it with the analog beam selection matrix, the result of which is equivalent to RF analog precoding. The system structure depicted in Fig. 7.2 can be regarded as a hybrid analog-digital precoding structure, and the RF analog precoding can be given by

$$\mathbf{P}_{\mathrm{RF}} = \mathbf{A}^H \mathbf{S}_{\mathrm{RF}} \in \mathbb{C}^{N_{\mathrm{T}} \times N_{\mathrm{RF}}}. \tag{7.17}$$

Therefore, we can use the hybrid analog-digital precoding design method to solve the above problems. To do so, the fully-digital baseband precoding matrix $\mathbf{P}_{\mathrm{D}} \in \mathbb{C}^{N_{\mathrm{T}} \times K}$ is first supposed to be given, and the received signal of satellites can be expressed as

$$\mathbf{y}^{'} = \mathbf{H}^H \mathbf{P}_{\mathrm{D}} \mathbf{s} + \mathbf{n}, \tag{7.18}$$

where $\mathbf{P}_{\mathrm{D}}$ is equivalent to $\mathbf{P}_{\mathrm{RF}}\mathbf{P}$, and is not constrained by the form of the analog precoding matrix.

In the uplink multi-satellite MIMO communication system, zero-forcing precoding is considered to be an asymptotically optimal linear precoding method in the sense that it provides an optimal precoding as the number of satellites increases without bound, which achieves high data rate [25, 33] when the CSI is known.[2] So we have

$$\mathbf{P}_{\mathrm{D}} = \mathbf{H}\left(\mathbf{H}^H \mathbf{H}\right)^{-1} \mathbf{D} \in \mathbb{C}^{N_{\mathrm{T}} \times K}, \tag{7.19}$$

where $\mathbf{D} \in \mathbb{C}^{K \times K}$ and the kth diagonal element is $\sqrt{d_k}$, $d_k > 0$. At this point, the data rate of the kth satellite can be expressed as

$$R_{\mathrm{ZF},k} = \log_2 \left(1 + \frac{d_k}{\sigma_n^2}\right). \tag{7.20}$$

The signal power sent to the kth satellite is

$$P_k^{'} = \left\|\mathbf{p}_{\mathrm{D},k} s_k\right\|^2 = \left\|\mathbf{p}_{\mathrm{D},k}\right\|^2 = d_k \left[\left(\mathbf{H}^H \mathbf{H}\right)^{-1}\right]_{k,k}, \tag{7.21}$$

where $\mathbf{p}_{\mathrm{D},k}$ denotes the kth column of $\mathbf{P}_{\mathrm{D}}$. At the same time, the total power consumption of the system still follows the lens antenna array power consumption model, and can be given by

$$P_{\mathrm{ZF,total}} = \frac{1}{\alpha} \sum_{k=1}^{K} d_k \left[\left(\mathbf{H}^H \mathbf{H}\right)^{-1}\right]_{k,k} + N_{\mathrm{RF}}(P_{\mathrm{RF}} + N_{\mathrm{T}} P_{\mathrm{SW}}) + P_{\mathrm{C}}. \tag{7.22}$$

[2] The results provide an upper-bound for the realistic case, where the CSI is only partially available.

Accordingly, the EE optimization problem can be restated as (7.23).

$$\mathcal{P}3: \max_{\mathbf{d}} \quad EE_{\mathrm{ZF}} = \frac{\sum_{k=1}^{K} R_{\mathrm{ZF},k}}{P_{\mathrm{ZF,total}}}$$

$$= \frac{\sum_{k=1}^{K} \log_2\left(1+\frac{d_k}{\sigma_n^2}\right)}{\frac{1}{\alpha}\sum_{k=1}^{K} d_k\left[\left(\mathbf{H}^H\mathbf{H}\right)^{-1}\right]_{k,k} + N_{\mathrm{RF}}(P_{\mathrm{RF}}+N_{\mathrm{T}}P_{\mathrm{SW}})+P_{\mathrm{C}}} \tag{7.23}$$

$$\text{s.t.} \quad \sum_{k=1}^{K} P_k = \sum_{k=1}^{K} d_k\left[\left(\mathbf{H}^H\mathbf{H}\right)^{-1}\right]_{k,k} \leqslant P_{\max}, \tag{7.23a}$$

$$d_k \geqslant 0, \tag{7.23b}$$

$$\gamma_k = \frac{d_k}{\sigma_n^2} \geqslant \gamma_{\min}. \tag{7.23c}$$

7.3.2 Fully-Digital Precoding Optimization

Problems ($\mathcal{P}3$) and ($\mathcal{P}2$) have the same form of objective function. By optimizing the design $\mathbf{d} = [d_1 \cdots d_K]$, the EE of (7.23) is maximized. It is only limited by the power constraints and the minimum SINR constraints. Therefore, the optimal solution of problem ($\mathcal{P}3$), that is, the maximum EE $EE_{\mathrm{ZF}}^{\max}$, is greater than or equal to the optimal solution $EE^{\max}$ of problem ($\mathcal{P}2$). In addition, the optimization problem is a fractional programming problem of concave function maximization, which can be solved by the Dinkelbach's algorithm. Following the Dinkelbach's algorithm, we define

$$f(\mathbf{d}) = \sum_{k=1}^{K} \log_2\left(1+\frac{d_k}{\sigma_n^2}\right), \tag{7.24}$$

$$g(\mathbf{d}) = \frac{1}{\alpha}\sum_{k=1}^{K} d_k\left[\left(\mathbf{H}^H\mathbf{H}\right)^{-1}\right]_{k,k} + N_{\mathrm{RF}}(P_{\mathrm{RF}}+N_{\mathrm{T}}P_{\mathrm{SW}})+P_{\mathrm{C}}. \tag{7.25}$$

Then, the optimization problem ($\mathcal{P}3$) can be reduced into the following optimization problem:

$$\begin{aligned} \mathcal{P}4: F(\eta) &= \max_{\mathbf{d}} \{f(\mathbf{d}) - \eta g(\mathbf{d})\} \\ \text{s.t.} &\quad (7.23\text{a}), (7.23\text{b}), (7.23\text{c}), \end{aligned} \tag{7.26}$$

where its optimal solution is obtained at $F(\eta) = 0$. The solution process can be outlined as follows.

7.3.2.1 Optimizing Variables $\mathbf{d} = [d_1 \cdots d_K]$

Given η, the optimization problem can be expressed as

$$\mathcal{P}5: \max_{d_k, k=1,\ldots,K, d_k \geqslant 0} \sum_{k=1}^{K} R_{\mathrm{ZF},k} - \eta P_{\mathrm{ZF,total}} \tag{7.27}$$

$$\text{s.t.} \quad \sum_{k=1}^{K} P_k^{'} \leqslant P_{\max}, \tag{7.27a}$$

$$\gamma_k \geqslant \gamma_{\min}. \tag{7.27b}$$

This problem is a standard concave function maximization problem. We rewrite it as the following convex optimization problem

$$\mathcal{P}6: \min_{d_k, k=1,\ldots,K, d_k \geqslant 0} -(f(\mathbf{d}) - \eta g(\mathbf{d})) \tag{7.28}$$

$$\text{s.t.} \quad \sum_{k=1}^{K} d_k w_k - P_{\max} \leqslant 0, \tag{7.28a}$$

$$\gamma_{\min} - \frac{d_k}{\sigma_n^2} \leqslant 0, \tag{7.28b}$$

where, $w_k = \left[\left(\mathbf{H}^H \mathbf{H}\right)^{-1}\right]_{k,k}$, $k = 1, \ldots, K$. In order to obtain the optimal solution of the above problem, we use the Lagrange multiplier method and introduce the Lagrange coefficients $\mu \geqslant 0$ and $\lambda_k \geqslant 0$, $k = 1, \ldots, K$. Then the Lagrange function can be expressed as

$$\begin{aligned} L(d_k, \mu, \lambda_k) &= -(f(\mathbf{d}) - \eta g(\mathbf{d})) \\ &+\mu \left(\sum_{k=1}^{K} d_k w_k - P_{\max} \right) + \sum_{k=1}^{K} \lambda_k \left(\gamma_{\min} - \frac{d_k}{\sigma_n^2} \right), \end{aligned} \tag{7.29}$$

and the KKT conditions can be given by

Algorithm 7.1 Fully-digital precoding matrix optimization

Initialization: Initialize the iteration counter $i = 0$ and the optimization accuracy ε (or the maximum number of iterations $I_{\max}$), $\eta^{(0)} = 0$

repeat
  $i = i + 1$
  given $\eta^{(i-1)}$, solve the problem (7.27) to get $\mathbf{d}^{(i)}$
  $F(\eta^{(i-1)}) = f(\mathbf{d}^{(i)}) - \eta^{(i-1)} g(\mathbf{d}^{(i)})$
  update $\eta^{(i)} = \frac{f(\mathbf{d}^{(i)})}{g(\mathbf{d}^{(i)})}$

until $\left|F(\eta^{(i-1)})\right| < \varepsilon$ (or $i = I_{\max}$)

Ensure: Optimal fully-digital precoding matrix $\mathbf{P}_{\mathrm{D}} = \mathbf{H}\left(\mathbf{H}^H\mathbf{H}\right)^{-1}\mathbf{D}^*$, where $\mathbf{D}^* = \mathrm{diag}\{\mathbf{d}^{(i)}\}$.

$$\begin{cases} \frac{\partial L(d_k, \mu, \lambda_k)}{d_k} = 0, \\ \mu\left(\sum\limits_{k=1}^{K} d_k w_k - P_{\max}\right) = 0, \lambda_k\left(\gamma_{\min} - \frac{d_k}{\sigma_n^2}\right) = 0, \\ \sum\limits_{k=1}^{K} d_k w_k \leqslant P_{\max}, \frac{d_k}{\sigma_n^2} \geqslant \gamma_{\min}, \\ \mu \geqslant 0, \lambda_k \geqslant 0, k = 1, \ldots, K. \end{cases} \tag{7.30}$$

Then, the optimal value of d_k can be given by

$$d_k^* = \max\left\{\frac{1}{\ln 2\left(\frac{\eta}{\alpha} + \mu\right) w_k} - \sigma_n^2, \sigma_n^2 \gamma_{\min}\right\}, k = 1, \ldots, K, \tag{7.31}$$

where the value of μ is chosen to meet the power constraints.

7.3.2.2 Updating η

We denote the value of η in the ith iteration by $\eta^{(i)}$, and the optimal d^* obtained in the ith iteration by $d^{(i)}$. Based on Dinkelbach's algorithm, η can be updated by the following formula:

$$\eta^{(i)} = \frac{f(\mathbf{d}^{(i)})}{g(\mathbf{d}^{(i)})}. \tag{7.32}$$

Based on the above discussion, the optimization of fully-digital precoding matrix is described in Algorithm 7.1.

7.3.3 Hybrid Analog-Digital Precoding Approximation

In order to make the performance of the hybrid analog-digital precoding close enough to the performance of the fully-digital precoding, we need to make the designed hybrid precoding $\mathbf{P}_{\mathrm{RF}}\mathbf{P}$ close enough to the fully-digital precoding matrix $\mathbf{P}_{\mathrm{D}}$. So we can consider the Euclidean distance as an approximate criterion to design the hybrid analog-digital precoding matrix [34, 35]. Specifically, we formulate the following problem.

$$\mathcal{P}7 : \left(\mathbf{P}_{\mathrm{RF}}^{opt}, \mathbf{P}^{opt}\right) = \underset{\mathbf{P}_{\mathrm{RF}},\mathbf{P}}{\arg\min} \quad \|\mathbf{P}_{\mathrm{D}} - \mathbf{P}_{\mathrm{RF}}\mathbf{P}\|_F^2 \tag{7.33}$$

$$\text{s.t.} \quad \mathbf{p}_{\mathrm{RF,n}} \in \{\mathbf{a}(\varphi_n), n = 1, \ldots, N_{\mathrm{T}}\}, \tag{7.33a}$$

$$\mathbf{p}_{\mathrm{RF,n}} \neq \mathbf{p}_{\mathrm{RF,m}}, \forall m \neq n, m, n = 1, \ldots, N_{\mathrm{RF}}, \tag{7.33b}$$

$$\|\mathbf{P}_{\mathrm{RF}}\mathbf{P}\|_F^2 = \|\mathbf{P}_{\mathrm{D}}\|_F^2, \tag{7.33c}$$

where $\mathbf{p}_{\mathrm{RF,n}}$ denotes the nth column of $\mathbf{P}_{\mathrm{RF,n}}$. As can be seen from the problem ($\mathcal{P}7$), the hybrid analog-digital precoding design is to select N_{RF} beam vector from the lens antenna array response matrix to construct an analog precoding matrix. Then, we find the corresponding baseband digital precoding matrix. Therefore, the problem ($\mathcal{P}7$) can be transformed into an equivalent problem as

$$\mathcal{P}8 : \tilde{\mathbf{P}}^{opt} = \underset{\tilde{\mathbf{P}}}{\arg\min} \quad \left\|\mathbf{P}_{\mathrm{D}} - \mathbf{A}^H\tilde{\mathbf{P}}\right\|_F^2 \tag{7.34}$$

$$\text{s.t.} \quad \left\|\mathrm{diag}\left(\tilde{\mathbf{P}}\tilde{\mathbf{P}}^H\right)\right\|_0 = N_{\mathrm{RF}}, \tag{7.34a}$$

$$\left\|\mathbf{A}^H\tilde{\mathbf{P}}\right\|_F^2 = \|\mathbf{P}_{\mathrm{D}}\|_F^2, \tag{7.34b}$$

where $\tilde{\mathbf{P}}$ is a $N_T \times K$ matrix. The sparsity constraint $\left\|\mathrm{diag}\left(\tilde{\mathbf{P}}\tilde{\mathbf{P}}^H\right)\right\|_0 = N_{\mathrm{RF}}$ represents that $\tilde{\mathbf{P}}$ has only N_{RF} non-zero rows. It means only the corresponding N_{RF} columns in $\mathbf{A}^H$ are selected. Then, the baseband digital precoding matrix $\mathbf{P}^{opt}$ can be obtained from N_{RF} non-zero rows of $\tilde{\mathbf{P}}^{opt}$, and we can also get the selection matrix $\mathbf{S}_{\mathrm{RF}}^{opt}$. Therefore, the optimization problem ($\mathcal{P}8$) can be solved by the method of sparse reconstruction [36, 37] as described in Algorithm 7.2.

Algorithm 7.2 Joint design of analog beam selection matrix $\mathbf{S}_{\text{RF}}$ and baseband digital precoding matrix $\mathbf{P}$

Initialization: Set analog beam selection matrix $\mathbf{S}_{\text{RF}}$ and matrix $\mathbf{p}_{\text{RF}}$ as empty matrix, $\mathbf{P}_{\text{res}} = \mathbf{P}_{\text{D}}$

For $i = 1 : N_{\text{RF}}$ **do**

$\mathbf{Q} = \mathbf{A}\mathbf{P}_{\text{res}}$

$n = \underset{l=1,\ldots,N_{\text{T}}}{\arg\max} \left[\mathbf{Q}\mathbf{Q}^H\right]_{l,l}$

$\mathbf{P}_{\text{RF}} = [\mathbf{P}_{\text{RF}}|\mathbf{a}(\varphi_n)]$

$\mathbf{S}_{\text{RF}} = [\mathbf{S}_{\text{RF}}|\mathbf{e}_n]$

$\mathbf{P} = \left(\mathbf{P}_{\text{RF}}^H \mathbf{P}_{\text{RF}}\right)^{-1} \mathbf{P}_{\text{RF}}^H \mathbf{P}_{\text{D}}$

$\mathbf{P}_{\text{res}} = \frac{\mathbf{P}_{\text{D}} - \mathbf{P}_{\text{RF}}\mathbf{P}}{\|\mathbf{P}_{\text{D}} - \mathbf{P}_{\text{RF}}\mathbf{P}\|_F}$

End for

$\mathbf{P} = \|\mathbf{P}_{\text{D}}\|_F \frac{\mathbf{P}}{\|\mathbf{P}_{\text{RF}}\mathbf{P}\|_F}$

Return $\mathbf{S}_{\text{RF}}, \mathbf{P}$

Algorithm 7.3 Optimal design of the number of required RF chains N_{RF}

Initialization: Set original number of required RF chains as a random integer N_{RF}, δ=0

repeat

For $i = 1 : L_{\text{M}}$ **do**

given N_{RF}, solve the problem ($\mathcal{P}3$) with Algorithm 1 and Algorithm 2 to get EE_0

get $N_{\text{RF}}^{'}$ randomly

$(\max(K, N_{\text{RF}} - N_\delta) \leq N_{\text{RF}}^{'} \leq \min(N_{\text{T}}, N_{\text{RF}} + N_\delta))$

given $N_{\text{RF}}^{'}$, solve the problem ($\mathcal{P}3$) with Algorithm 1 and Algorithm 2 to get $EE^{'}$

$\delta = EE^{'} - EE_0$

If $\delta > 0 | e^{\delta/T} > rand$

$EE_0 = EE^{'}$

$N_{\text{RF}} = N_{\text{RF}}^{'}$

$T_0 = \xi * T_0$

until $T_0 < T_{\text{end}}$

Return N_{RF}, EE_0

7.4 Energy-Efficient RF Link Number Design Methodology

Based on Algorithms 7.1 and 7.2, we can obtain the analog beam selection matrix $\mathbf{S}_{\text{RF}}$ and baseband digital precoding matrix $\mathbf{P}$ aiming at maximizing EE with given number of required RF chains. In this section, we will further optimize the number of required RF chains to improve system EE.

The number of required RF chains N_{RF} affects the dimensions and elements of the analog beam selection matrix $\mathbf{S}_{\text{RF}}$ and the baseband digital precoding matrix $\mathbf{P}$. It is also related to system power consumption. So it is difficult to obtain an analytical solution for the optimal number of required RF chains. However, the number of

Table 7.1 System parameters for simulation

Parameter	Values
Maximum signal transmit power $P_{\max}$ (dBW)	−10–10
# of antennas N_{T}	21–201
# of selected RF chains N_{RF}	3–21
# of connected satellites K	3–12

required RF chains N_{RF} is a positive integer, and $K \le N_{\mathrm{RF}} \le N_{\mathrm{T}}$. Simulated annealing algorithm (SAA) can be considered to solve the nonlinear integer optimization problem, searching for an optimal number of required RF chains efficiently [38]. SAA is a global optimal algorithm based on the similarity between the solution of the optimization problem and the annealing process of the physical system, which has higher computational efficiency compared to the enumeration method and the genetic algorithm [39], as well as more accurate results compared to the greedy algorithm [40]. It uses the Metropolis algorithm and appropriately controls the temperature drop process to achieve simulated annealing, so as to complete the goal of solving the problem ($\mathcal{P}1$), finding the most suitable integer solution within the given range in a short time by adjusting the appropriate values of parameters. In this case, to be specific, the initial temperature is set to be T_0, the lower bound of temperature is set to be T_{end}, and the temperature decreases by attenuation coefficient, a certain ratio ξ in each loop, which is only terminated when the temperature reaches the lower bound. The length of the Markov chain is set to be L_{M} based on multiple parameters such as attenuation coefficient and initial solution, and after several practical trials and comparisons, so there will be L_{M} disturbances at each temperature. A random value of the number of required RF chains N_{RF} is also initialized. In each iteration, a new value $N_{\mathrm{RF}}^{'}$ within the distance of N_{δ} from the original N_{RF} is generated and examined by comparing the EE with the original one. The higher can be regarded as the new EE, and the lower has the chance of being regarded as the new EE with a specified probability. The method is described in Algorithm 7.3.

Combining Algorithms 7.1 with 7.2, we can find the number of required RF chains that maximizes EE and its corresponding analog beam selection matrix and baseband digital precoding matrix.

7.5 Numerical Results and Analysis

In this section, the proposed algorithm will be simulated and verified with the Monte Carlo simulation method, which is based on the Matlab simulation platform. The EE in the MIMO multi-satellite uplink communication system will also be illustrated. The key system parameters are shown in Table 7.1. Unless otherwise specified, the number of antennas is set to $N_{\mathrm{T}} = 21$ and the number of satellites is set to $K = 3$.

Table 7.2 Comparison of circuit power consumption between two structures

Array structure ($N_T = 21$)	Number of RF links coefficients	Circuit power consumption (W)
Fully-digital structure	$N_{RF} = N_T = 21$	15.25
Lens antenna array	$N_{RF} = 3$	11.065
	$N_{RF} = 6$	12.13
	$N_{RF} = 9$	13.195

The number of channel paths between the ground terminal and each satellite is set to 3, including 1 LoS path and 2 NLoS paths. In general, the LoS path gain is higher than the NLoS path gain, and there are $\beta_{k,0} \in \mathcal{CN}(0, 1)$, $\beta_{k,l} \in \mathcal{CN}(0, 10^{-1})$. The angles $\phi_{k,0}$ and $\phi_{k,l}$ associated with the paths are assumed to follow the uniform distribution within $[-\pi/2, \pi/2]$. In addition, for the circuit power consumption part, RF chain power consumption is $P_{RF} = 250$ mW, analog switch power consumption is $P_{SW} = 5$ mW, other circuit power consumption is $P_C = 10$ W, power amplifier efficiency is $\alpha = 0.3$. The number of trials in the following results is set to 1000.

We first discuss the impact of the use of the lens antenna array on the system performance. Table 7.2 shows the circuit power consumption comparison between the fully-digital structure and the lens antenna array structure. Since the number of required RF chains in the lens antenna array structure is small, its power consumption is lower than the fully-digital structure and increases with the number of required RF chains. In addition, Fig. 7.4 shows the spectral efficiency against the maximum signal transmit power P_{max} of the two structures. With the increase of the maximum signal transmission power, the spectral efficiency of the two structures is improved at the same time. Additionally, the spectral efficiency of a lens antenna array structure increases as the number of required RF chains increases. It can be seen from Fig. 7.4 that the performance is basically the same as that of the fully-digital structure at $N_{RF} = 9$, while 12 RF chains are reduced compared to the latter structure. Therefore, for the lens antenna array structure, combined with the energy-efficient precoding design, the optimal design of the number of required RF chains can achieve a compromise between system spectral efficiency and power consumption, which is the EE optimization problem studied in this chapter.

Figure 7.5 shows the EE against the maximum signal transmission power P_{max}. The optimal design method of the number of required RF chains for improving EE and the method of fixing the number of required RF chains are compared. The maximum signal transmission power P_{max} is set to −10–10 dBW as that in Fig. 7.4. As can be seen from Fig. 7.5, the EE increases with the increase of the maximum transmit power for all methods. Although increasing the number of required RF chains can improve the system spectral efficiency, it can be seen from Fig. 7.5 that there is no clear correspondence between the system EE and the number of required RF chains. This is because the number of required RF chains also affects the system power consumption, which also shows the necessity of optimizing the number of required RF chains. The simulation results show that the EE performance of the

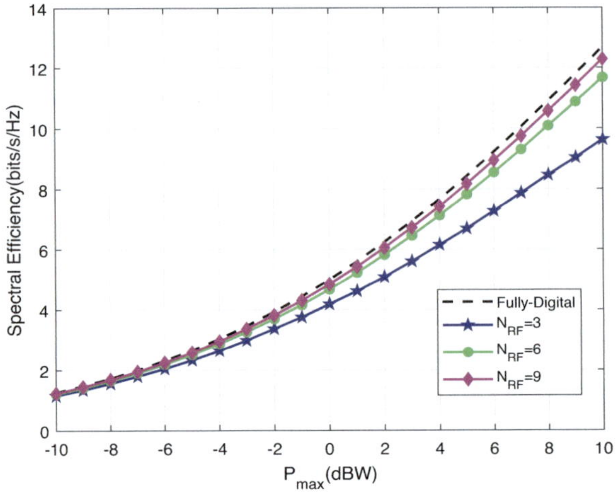

Fig. 7.4 Spectral efficiency versus maximum transmit power for fully-digital and fixed number of RF chains N_{RF} lens antenna array structure, $N_{\mathrm{T}} = 21$, $K = 3$, $L = 3$, $N_{\mathrm{RF}} = 3$

Fig. 7.5 EE versus maximum transmit power for fully-digital structure, N_{RF} number optimization design and fixed N_{RF} number method, $N_{\mathrm{T}} = 21$, $K = 3$, $L = 3$

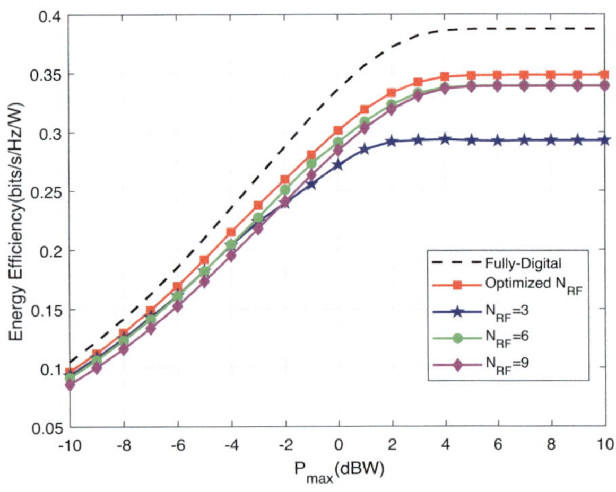

optimal design method for improving the number of energy-efficient RF chains in this chapter is always better than the method for fixing the number of required RF chains, and the EE values become closer to the fully-digital structure after using the optimal design on the number of required RF chains.

Figure 7.6 shows the EE against the number of antennas N_{T} with respect to the optimal design method for the number of energy-efficient RF chains and the method for fixing the number of required RF chains in this chapter, where $P_{\max} = 0$ dBW. It can be seen from the figure that the EE of both methods increases with the number of antennas from 21 to 201. This is because the increase of the number of antennas will improve the spatial resolution of the lens antenna array, thereby reducing the

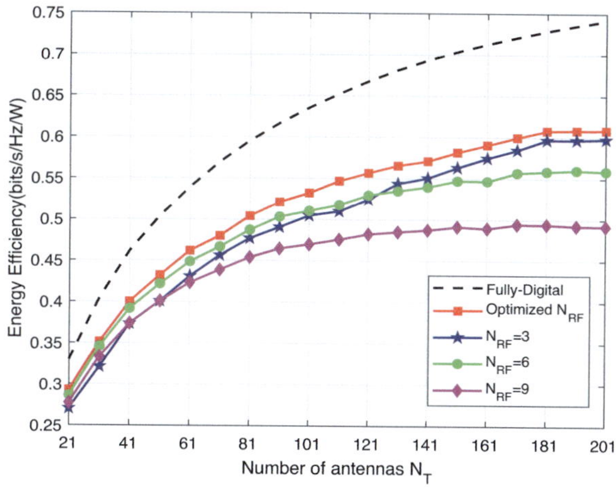

Fig. 7.6 EE versus number of antennas N_{T} for fully-digital structure, N_{RF} number optimization design and fixed N_{RF} number method, $K = 3$, $L = 3$

interference between satellites. In addition, as the number of antennas increases to a certain value, the use of fewer RF chains can achieve better EE performance. As discussed in Sect. 7.2, the increase in spatial resolution makes the spatial direction gain of the corresponding satellite more concentrated on certain antennas. Moreover, the LoS path gain is higher than the NLoS path gain in this chapter, and the main space direction gain on the corresponding satellite will be concentrated on fewer antennas. Therefore, the communication can be completed with a smaller number of required RF chains without significant performance loss. Fewer RF chains means lower system power consumption, resulting in better EE performance. As can be seen from Figs. 7.5 and 7.6, the method proposed in this chapter can dynamically optimize the design of the number of required RF chains as well as their corresponding analog beam selection matrix and baseband digital precoding matrix according to the system parameters, so as to obtain the optimal EE performance. The EE gap under different numbers of RF chains increases with the increase of the maximum transmit power and the number of antennas. So, optimizing the number of RF chains acquires more gains when the maximum transmit power or the number of antennas is relatively large.

As can be seen from the previous discussion, the relative positions of satellites have an important impact on system performance. In order to better study this problem, we set $K = 2$, where K is the number of satellites in the transmission system, and there are only LoS paths. Therefore, we define the spatial directions of the satellite 1 and satellite 2 as θ_1 and θ_2. We obtain the EE performance curve in Fig. 7.7 by fixing the value of θ_1, and changing θ_2 to increase the distance between θ_1 and θ_2, i.e., $\Delta\theta = \theta_2 - \theta_1$. It can be seen from the figure that with the increase of $\Delta\theta$, which is set to 1/210 degrees in the simulation, the spatial direction difference between two satellites increases accordingly, and the EE of the system is continuously improved. It is because the interference between the beams of the lens antenna array is reduced, thereby reducing the interference between satellites. As $\Delta\theta$ continues to increase,

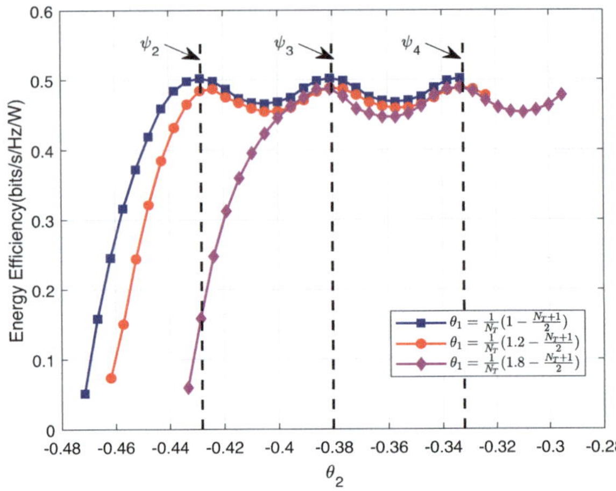

Fig. 7.7 EE versus satellites spatial direction for different angle distances from the preset direction of lens antenna array in the case of LoS path, $N_{\mathrm{T}} = 21, K = 2, L = 1$

the system EE will fluctuate at a higher level. It is because the inter-beam interference is small when the value of $\Delta\theta$ is large, and fluctuates with the change of $\Delta\theta$ at a lower level, which can be seen from the beamspace response in Fig. 7.3. In addition, it can be seen from Fig. 7.7 that when the value of $\Delta\theta$ is larger and the value of θ_2 is consistent with the preset direction of the lens antenna array, the EE curve will reach the maximum value. It is because the beam corresponding to satellite 2 will not interfere with the beam of satellite 1 and the interference it gets from the beam of satellite 1 is relatively small at the same time. In particular, when θ_1 is consistent with the preset direction of the lens antenna array, which is the blue curve in the figure, the beam corresponding to satellite 1 will not interfere with the beam corresponding to satellite 2, and the system will obtain a EE performance better than that when we chose other spatial directions of satellite 1. From the value range of the system EE curve in Fig. 7.7, it can be seen that when θ_2 is greater than half of the spatial direction resolution of the lens antenna array, which is $1/(2N_{\mathrm{T}})$, the system obtains better EE performance. Therefore, in the deployment for the ground terminal and satellites in the transmission system, we can jointly design the number of antennas of the lens antenna array and the relative position of satellites to meet the system performance requirements.

Figure 7.8 shows the system EE against the number of required RF chains when the system simulation parameters are consistent with those in Fig. 7.4. It can be seen from the figure that with the increase of the maximum transmit power, the system EE increases as well. Also, when the number of required RF chains selected increases, the system EE increases at first until the selected value reaches the optimal number of required RF chains produced by the SAA, then gradually decreases. This is because before the optimal number of required RF chains is reached, the increase of the number of required RF chains can improve the transmitting rate, and this influence is greater than the extra power consumption brought by the increasing number of chains. When the number of required RF chains continues to increase, the system

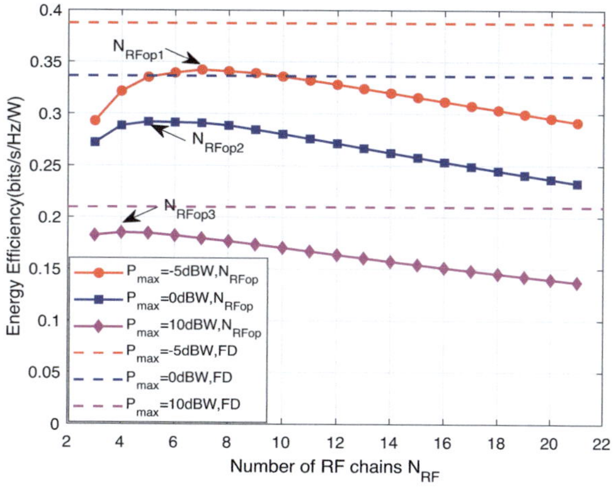

Fig. 7.8 EE versus the number of required RF chains N_{RF} for fully-digital structure and N_{RF} number optimization design method for maximum transmit power set to be -5, 0 and 10 dBW, $N_T = 21$, $K = 3$, $L = 3$

power consumption will increase accordingly, which is non-negligible compared to the transmitting rate, resulting in a relative decline in system performance. We can also find that with the increase of maximum transmit power, the ground terminal is more likely to select proper or larger number of RF chains through the number of RF chain optimization algorithms to achieve better system performance, so the optimal number of required RF chains becomes relatively larger.

Figure 7.9 shows the EE against the number of satellites K, which is set to 3–9. From the theoretical analysis and simulation results, the number of required RF chains N_{RF} is usually larger than or equal to that of the satellites. In the simulation, with the increase of K, the value of EE increases accordingly. Moreover, the increasing trend is relatively obvious when K is small. Then, as K grows larger, the value of

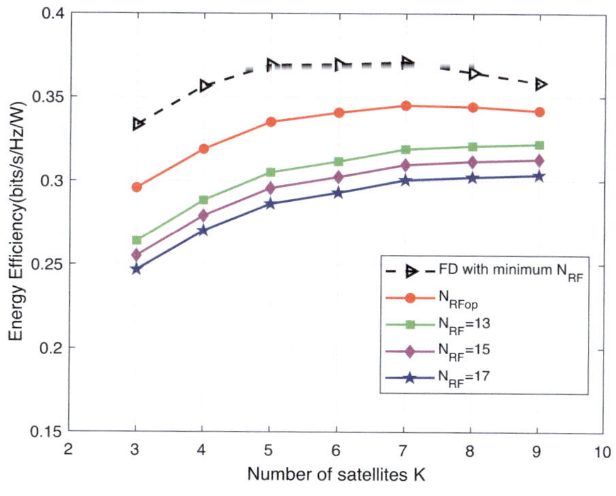

Fig. 7.9 EE versus satellite number K for fully-digital structure, N_{RF} number optimization design and fixed N_{RF} number method, $N_T = 21$, $L = 3$

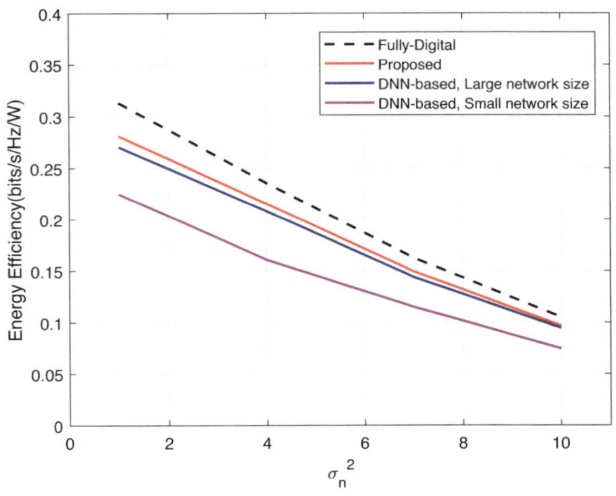

Fig. 7.10 EE versus power of noise for a fully-digital structure, proposed method and machine learning-based precoding optimization design

EE fluctuates at a higher level with a small decreasing trend. The EE value of fully-digital structure becomes closer to the one of N_{RF} number optimization design. This is because a larger number of satellites needs a larger transmitting sum rate, while the continuously growing number of satellites increases the power needed and reduces the spatial direction difference. Besides, when K is small, the optimal number of N_{RF} obtained by simulation may be larger than K. However, when K grows larger, the optimal N_{RF} obtained is likely to equal to K. Compared to the results produced by fixed numbers of RF chains, the designed optimal number of required RF chains can obtain better system performance.

Figure 7.10 shows that the proposed method has better performance in optimizing EE than the machine learning-based precoding optimization method with a small network size. It has similar performance compared with the machine learning-based precoding optimization method with a large network size, while the latter has higher complexity.

7.6 Conclusions

This chapter studies the EE optimization problem of MIMO multi-satellite uplink communication systems based on lens antenna arrays. Firstly, we establish the signal model of the communication system based on the lens antenna array, and analyze the characteristics of the beamspace channel. Then, we formulate an EE optimization problem that jointly optimizes the number of required RF chains, the analog beam selection matrix and the baseband digital precoding matrix. In order to solve this non-convex problem, we first optimize the analog beam selection and baseband digital precoding under a given number of required RF chains based on zero forcing-based precoding. Then, we optimize the number of required RF chains based on the

SAA, which can dynamically select the number of required RF chains according to the system and channel parameters, so as to maximize the EE of the system. Finally, we compare the performance of the lens antenna array and the traditional fully-digital structure by simulation, which verifies that the EE performance of the proposed optimization scheme is better than that of the fixed number of RF chains. In addition, we also discuss the influence of the relative positions between satellites on the system performance. The simulation results show that when the spatial direction difference of the satellites is greater than half of the spatial direction resolution of the lens antenna array, the system can obtain better performance, which has a certain influence on the deployment of MIMO systems based on the lens antenna array.

References

1. Cao Y, Yan S (2021) Multi-beam SIW leaky-wave antenna with 2-D beam scanning capability for millimeter-wave radar applications. Int J RF Microw Comput Aided Eng 31(5)
2. Garg R, Natarajan AS (2017) A 28-GHz low-power phased-array receiver front-end with 360° RTPS phase shift range. IEEE Trans Microw Theory Tech 65(11):4703–4714
3. Cao Y, Yan S (2021) A low-profile high-gain multi-beam antenna based on cylindrical metasurface Luneburg lens. In: Computing, Communication, and IoT Applications. IEEE, pp 35–38
4. Matytsine L, Lagoiski P, Matytsine M et al (2012) Large size, lightweight, Luneburg lenses for multi-beam antenna applications. In: European conference on antennas and propagation. IEEE, pp 2266–2270
5. Yang L, Zeng Y, Zhang R (2016) In-band wireless information and power transfer with lens antenna array. IEEE Commun Lett 21(1):100–103
6. Wang H-F, Wang Z-B, Wu Z-H et al (2019) Beam-scanning lens antenna based on elliptical paraboloid phase distribution metasurfaces. IEEE Antennas Wireless Propag. Lett. 18(8):1562–1566
7. Thakur SK, Rukmini TS, Ravishankar S et al (2019) Collimation properties of microstrip patch fed dielectric lens antenna for broadband mobile communication. In: International Conference on Future Computer and Communication. IEEE, pp 522–526
8. Thornton J, Butlin T, Dalay B (2012) Lens-reflector array antenna for satellite communications on the move. In: European Space Agency Workshop Satcom User Terminal Antennas, pp 3–5
9. Schoenlinner B, Wu X, Ebling JP et al (2002) Wide-scan spherical-lens antennas for automotive radars. IEEE Trans Microw Theory Techn 50(9):2166–2175
10. Afanasyev P, Matitsine S, Sledkov V et al (2015) Multi-beam Luneburg lens antenna for cellular communications. In: European conference on antennas and propagation. IEEE, pp 1–4
11. Gong S, Wang S, Chen S et al (2019) Robust energy efficiency optimization for amplify-and-forward MIMO relaying systems. IEEE Trans Wireless Commun 18(9):4326–4343
12. Wu Q, Tao M, Kwan Ng DW et al (2016) Energy-efficient resource allocation for wireless powered communication networks. IEEE Trans Wireless Commun 15(3):2312–2327
13. Yuan W, Wei Z, Li S et al (2021) Integrated sensing and communication-assisted orthogonal time frequency space transmission for vehicular networks. IEEE J Sel Topics Signal Process 15(6):1515–1528
14. Tang S, Ma Z, Xiao M et al (2020) Hybrid transceiver design for beamspace MIMO-NOMA in code-domain for mmWave communication using lens antenna array. IEEE J Sel Areas Commun 38(9):2118–2127

15. Zeng Y, Yang L, Zhang R (2018) Multi-user millimeter wave MIMO with full-dimensional lens antenna array. IEEE Trans Wireless Commun 17(4):2800–2814
16. Hu Q, Liu Y, Cai Y et al (2021) Joint deep reinforcement learning and unfolding: Beam selection and precoding for mmWave multiuser MIMO with lens arrays. IEEE J Sel Areas Commun 39(8):2289–2304
17. Pan J, Ye N, Yu H et al (2022) AI-driven blind signature classification for IoT connectivity: A deep learning approach. IEEE Trans Wireless Commun 21(8):6033–6047
18. Wang Y, Lu H, Zhao D et al (2020) Energy efficiency optimization in IRS-enhanced mmWave systems with lens antenna array. In: IEEE Global Communications Conference. IEEE, pp 1–6
19. Wang B, Dai L, Wang Z et al (2017) Spectrum and energy-efficient beamspace MIMO-NOMA for millimeter-wave communications using lens antenna array. IEEE J Sel Areas Commun 35(10):2370–2382
20. Shen W, Bu X, Gao X et al (2019) Beamspace precoding and beam selection for wide-band millimeter-wave MIMO relying on lens antenna arrays. IEEE Trans Signal Process 67(24):6301–6313
21. Gao X, Dai L, Sun Y et al (2017) Machine learning inspired energy efficient hybrid precoding for mmWave massive MIMO systems. In: IEEE International Conference on Communications. IEEE, pp 1–6
22. Ding T, Zhao Y, Li L et al (2019) Hybrid precoding for beamspace MIMO systems with sub-connected switches: a machine learning approach. IEEE Access 7:143273–143281
23. Kossey MR, Rizk C, Foster AC (2018) End-fire silicon optical phased array with half-wavelength spacing. APL Photonics 3(1)
24. Roh W, Seol J-Y, Park J et al (2014) Millimeter-wave beamforming as an enabling technology for 5G cellular communications: theoretical feasibility and prototype results. IEEE Commun Mag 52(2):106–113
25. Rusek F, Persson D, Lau BK et al (2013) Scaling up MIMO: opportunities and challenges with very large arrays. IEEE Signal Process Mag 30(1):40–60
26. Brady J, Behdad N, Sayeed AM (2013) Beamspace MIMO for millimeter-wave communications: system architecture, modeling, analysis, and measurements. IEEE Trans Antennas Propag 61(7):3814–3827
27. Zhang Y, Fusco V (2012) Fourier transform using a Rotman lens. In: European Microwave Conference. IEEE, pp 715–718
28. Lian J-W, Ban Y-L, Zhu H et al (2020) Reduced-sidelobe multibeam array antenna based on SIW Rotman lens. IEEE Antennas Wireless Propag Lett 19(1):188–192
29. Zeng Y, Zhang R (2016) Millimeter wave MIMO with lens antenna array: a new path division multiplexing paradigm. IEEE Trans Commun 64(4):1557–1571
30. Rangan S, Rappaport TS, Erkip E (2014) Millimeter-wave cellular wireless networks: potentials and challenges. In: IEEE, pp 366–385
31. Giordani M, Zorzi M (2020) Satellite communication at millimeter waves: a key enabler of the 6G era. In: International conference on computing, networking and communications. IEEE, pp 383–388
32. Méndez-Rial R, Rusu C, González-Prelcic N et al (2016) Hybrid MIMO architectures for millimeter wave communications: phase shifters or switches. IEEE Access 4:247–267
33. Boccardi F, Huang H (2006) Zero-forcing precoding for the MIMO broadcast channel under per-antenna power constraints. In: IEEE workshop signal process. Advanced Wireless Communications. IEEE, pp 1–5
34. Ayach OE, Rajagopal S, Abu-Surra S et al (2014) Spatially sparse precoding in millimeter wave MIMO systems. IEEE Trans Wireless Commun 13(3):1499–1513
35. Gao X, Dai L, Han S et al (2017) Reliable beamspace channel estimation for millimeter-wave massive MIMO systems with lens antenna array. IEEE Trans Wireless Commun 16(9):6010–6021
36. Tropp JA, Gilbert AC (2007) Signal recovery from random measurements via orthogonal matching pursuit. IEEE Trans Inf Theory 53(12):4655–4666

37. Rebollo-Neira L, Lowe D (2002) Optimized orthogonal matching pursuit approach. IEEE Signal Process Lett 9(4):137–140
38. Morales-Castañeda B, Zaldívar D, Cuevas E et al (2019) An improved simulated annealing algorithm based on ancient metallurgy techniques. Appl Soft Comput 84:40–60
39. Lambora A, Gupta K, Chopra K (2019) Genetic algorithm - a literature review. In: International Conference on Machine Learning, Big Data, Cloud and Parallel Computing. IEEE, pp 380–384
40. Lee S, Kim SB (2019) Parallel simulated annealing with a greedy algorithm for Bayesian network structure learning. IEEE Trans Knowl Data Eng 32(6):1157–1166

Chapter 8
Summary and Outlook

Kexin Zhou, Xuhui Ding, and Gaoyang Li

This chapter summarizes the book and discusses the future development direction of the MIMO communication system.

8.1 Summary

In this book, we discuss various technologies to improve the performance of networks and to achieve MIMO communications. The specific contributions of this book are as follows.

1. In Chap. 2, we have investigated the training optimizations for hybrid MIMO communications. Channel estimation is conceived for hybrid MIMO communication systems. Both mean square error minimization and MI maximization are used as our performance metrics, and a pair of low-complexity channel estimation schemes are proposed. In each scheme, the training sequence and the analog matrices of the transmitter and receiver are jointly optimized. We commence by designing the optimal training sequences and analog matrices for the first scheme. Upon relying on the resultant optimal structures, the training optimization problems are substantially simplified, and the non-convexity resulting from the analog matrices can be overcome. In the second scheme, the channel estimation and data transmission share the same analog matrices, which beneficially

K. Zhou · X. Ding
Beijing Institute of Technology, Beijing, China
e-mail: zhoukexin@bit.edu.cn

X. Ding
e-mail: dingxuhui@bit.edu.cn

G. Li (✉)
Beihang University, Beijing, China
e-mail: ligaoyang@buaa.edu.cn

X. Ding et al. (eds.), *Performance Analysis and Improvement in MIMO Communication Systems*, https://doi.org/10.1007/978-981-96-5951-7_8

reduces the overhead of optimizing the associated analog matrices. Therefore, a composite channel matrix is estimated instead of the true channel matrix. By exploiting the statistical optimization framework advocated, the analog matrices can be designed independently of the training sequence. Based on the resultant analog matrices, the training sequence can then be efficiently designed according to diverse channel statistics and performance metrics. Finally, we conclude by quantifying the performance benefits of the proposed estimation schemes.

2. In Chap. 3, we have investigated the problems of accurate time delay estimation, the design of training pilots, and hybrid matrix optimization within the large-scale antenna array THz broadband communication system. In contrast to the existing research based on narrow-band arrays, we hereby shed light on the time delay estimation of broadband arrays. In THz broadband communication systems, the data symbol duration is relatively short when compared with the dimension of the antenna array. In large-scale antenna systems, signals received in each antenna are no longer different phase-shifted copies of the same symbol but are completely different symbols, for which occasion, a traditional narrow-band structure is no longer suitable. Based on the above conclusion, we first proposed a system model based on large-scale antenna arrays and TDL structure. Secondly, we deduce the CRLB of the time delay estimation and present a time delay estimation algorithm that could reach the CRLB. Thirdly, by minimizing the CRLB, we address the design of the training pilot and optimize the TDL structure under the condition of a constant envelope training pilot and a modulus TDL structure. Finally, we disclose the numerical simulation results. According to the simulation results, the aforementioned method is workable in reaching the CRLB, the TDL structure can significantly surpass that of the traditional model, and the optimal pilot design method outperforms the pseudo-random pilot structure.
3. In Chap. 4, we have investigated the problem of AoA tracking for the large-scale array in THz communication, which has a large size and a narrow beam, highly demanding an accurate angle estimation. On the one hand, the system usually adopts a hybrid structure with limited RF chains, which increases the difficulty of angle estimation; on the other hand, the rapid mobility of users also brings new challenges to angle estimation. To address the above challenges, a two-stage tracking framework is proposed, which employs the random phase matrix and orthogonal long pilots in the first stage to reduce the complicated multi-user estimation to multiple single-user estimations, followed by using both wide and narrow beams in the second stage to serve high-speed and low-speed users. Furthermore, a GAMP method is proposed for facilitating a low-accuracy estimation of the angles, followed by adopting a modified EM algorithm-based phase estimation to unbiasedly estimate the instantaneous angle with the help of high-gain characteristics of the beams. The proposed structure can not only simplify the estimation complexity but also improve the estimation accuracy due to its capability of transferring the non-linear problem of angle observation into a linear Gaussian model. In addition, the Kalman tracking framework is employed to perform continuous angle tracking. Numerical results show that the

angle estimation based on the random phase matrix in the initial stage can obtain a high enough estimation accuracy, while the GAMP algorithm implemented in the second stage can quickly capture the angle range under the Rayleigh limit. The performance of the proposed EM-based tracking method is shown to outperform the traditional EKF method.

4. In Chap. 5, we investigate a joint angle-delay estimation in the THz communication system with a large-scale array. Due to the relatively high dimension of the THz array and the wideband of the signal, the traditional narrow-band-array assumption that signals received in different antennas are coherent cannot be satisfied. Meanwhile, high accuracy in angle-delay estimation could lead to a huge computation complexity. To deal with this contradiction, we first give out the system model of a large-scale array system and show the fact that signal bandwidth could be spread. After that, we present the CRLB of joint angle-delay estimation. Based on this system model, we prove that signals with different angle-delay are approximately orthogonal. Furthermore, two estimators are proposed: MLE with grid and PRCE without grid. To overcome the huge complexity, we provide a new compressed sensing structure with a sparsely uniform line array, random analog matrix, and WMC. A GAMP-based algorithm is listed to solve this CS problem. Simulation shows that the estimators proposed in this chapter outperform those based on the MUSIC method. In particular, MLE could reach CRLB but is limited by the grid, while the other one could reach CRLB at a high SNR. The GAMP-based algorithm jointly estimates angle-delay at the cost of a lower complexity, whose accuracy reaches the Rayleigh limit.
5. In Chap. 6, we have investigated in depth the two-timescale robust system EE precoding design for the multiuser dual-polarized MIMO system. To achieve good performance, low feedback overhead, and low implementation complexity, the dual-structured linear precoding scheme is adopted, which is based on the two timescale CSI and the dual-polarized antenna structure. The subgrouping technique, which is based on polarization, is also utilized to divide spatially grouped users into co-polarized subgroups to further reduce channel feedback overhead. The proposed robust system EE precoding design can achieve the maximization of the worst-case system EE with the norm-bounded channel errors of all users. This robust EE optimization is naturally decomposed into two parts. In the first part, based on the polarized spatial correlation information, the block diagonalization is utilized to design the preprocessing matrix. In the second part, based on the relationship between the mean square error and the signal-to-interference-plus-noise ratio, the linear precoding matrix can be optimized by utilizing the sign-definiteness lemma and the fractional programming technique. Specifically, the corresponding nonconvex EE fractional optimization problem is converted to a series of SDP problems, which are solved efficiently by the convex optimization method. Simulation results indicate that the proposed two timescale based dual-structured precoding has many advantages for robust system EE performance in the dual-polarized multiuser MIMO system.
6. In Chap. 7, we have investigated the EE optimization problem of MIMO multi-satellite uplink communication systems based on lens antenna arrays. With the

ever-increasing densification of low Earth orbit constellations, lens antenna array has become an attractive option for ground terminals to simultaneously connect to multiple satellites with improved coverage under insufficient energy supply. To this end, an optimization problem is formulated to maximize the EE of the system by respecting the uplink transmit power budget and the quality of service constraints. This problem has a discrete non-concave objective function and is solved with reduced computational complexity using zero forcing-based precoding and problem decomposition. The proposed algorithm iterates between the following three steps. In the first step, a zero forcing-based fully-digital precoder is first optimized under a fixed number of RF chains by adopting Dinkelbach's algorithm. Then, a hybrid analog-digital precoding based on analog beam selection and baseband digital precoding is designed to approximate the optimal fully-digital precoder under Euclidean distance criteria. On this basis, SAA is applied to further optimize the number of required RF chains with the aim of reducing total energy cost. Simulation results demonstrate that the proposed design achieves an improved performance in EE compared with existing schemes with a fixed number of RF chains by up to 15%.

8.2 Future Directions

In this section, we discuss the future directions of MIMO communication and its performance analysis.

8.2.1 UAV MIMO Systems

UAV platform demands high standards in terms of design complexity, cost, EE, and communication capacity, thus providing an ideal scenario for MIMO technology to fully exploit its advantages. As communication frequencies continue to rise, resulting in decreased antenna size, UAV platforms can now equip large-scale antenna arrays for MIMO communication. During missions, UAVs necessitate real-time large-scale data transmission, including video streams and sensor data. The diversity gain inherent in MIMO technology ensures the reliable transmission of the various data streams generated by UAVs. Additionally, within the MIMO field, beamforming technology offers UAV data links, spatial anti-interference, and spatial multiplexing capabilities. Furthermore, MIMO systems dynamically adjust communication parameters and optimize channel resources, thereby enhancing UAV networking flexibility. Research into MIMO communication technology based on UAV platforms holds great promise. In the future, popular research directions include: optimizing EE of UAV MIMO systems, improving placement optimization of UAV networks at the topological structure level, enhancing UAV MIMO systems' ability to resist electromagnetic interference signals, exploring Distributed MIMO technology for improved data transmission rates, coverage, and communication reliability, and integrating artificial intelligence technology to develop intelligent resource management and scheduling algorithms that adapt dynamically to complex and variable network environments.

8.2.2 MmWave-THz MIMO Systems

The advancement of 5G and future sixth-generation (6G) network technology has stimulated the surging demand for high-speed data transmission. In this context, mmWave-THz massive MIMO communication springs as an emerging technology critically enhancing both user throughput and spectral efficiency. Specifically, the mmWave ranging from 30 GHz to 300 GHz and the THz ranging from 0.1 THz to 10 THz are characterized by the expansive available bandwidth and short wavelength, facilitating high data rates and small antenna size for massive MIMO systems. Besides, the larger-scale antenna array offers significantly high beamforming gains to compensate for the severe path losses in the mmWave and THz communications. Nevertheless, despite the promising potential of mmWave-THz MIMO technology, its future research and technological advancement encounter significant challenges. On the one hand, the higher operating frequency imposes more stringent hardware requirements, particularly in the mmWave-THz frequency band, where the utilization of extremely small antenna components and dense antenna arrays presents substantial challenges for precision manufacturing and integration. Furthermore, the integration of a large number of antenna elements to achieve beamforming and spatial multiplexing makes the design of low-complexity hybrid precoding matrices complex, consequently driving the demand for sophisticated signal processing algorithms, higher computing resources, and real-time processing capabilities. In general, mmWave-THz massive MIMO technology provides a feasible path for future high-speed communication, and the exploration direction of low hardware cost and high processing EE is worth investigating.

8.2.3 AI MIMO Systems

With the development of MIMO technology enhancing data transmission rates, network capacity, and multi-user transmission capabilities, computational complexity has been significantly increased with large-scale implementations and the adoption of other crucial 5G enabling technologies. Traditional optimization methods have certain limitations and disadvantages, which may not favor extensive deployment. Machine learning, a subset of artificial intelligence (AI) techniques, has emerged as a promising solution over the past decade, capable of handling complex optimization problems. Machine learning algorithms, once properly trained, can achieve low complexity channel estimation and play a role in various optimization challenges related to wireless communications, including optimizing signal processing algorithms for MIMO systems, dynamically adjusting wireless spectrum resources, and enhancing communication quality and efficiency. Additionally, intelligent network systems can monitor and predict network states in real-time, proactively addressing potential issues, and thereby providing more stable and reliable communication services. Despite the absence of a standardized framework for integrating machine learning with MIMO, leveraging AI to achieve efficient, precise, and rapid algorithms holds significant developmental potential in the future deployment of 5G/6G networks.

8.2.4 Satellite Communication MIMO Systems

Satellite communication has become a cornerstone for expanding and supplementing terrestrial 5G networks due to its wide coverage and resilience to natural disasters. Consequently, research on satellite communication MIMO has been increasing. However, most of this research still focuses on geostationary orbit (GEO). Low earth orbit (LEO) satellites are considered a promising infrastructure, capable of providing high-throughput internet access to underserved or unserved areas with shorter propagation delays, less path loss, and lower launch costs. Given the expansion of the LEO market and the provision of actual services in recent years, there is a strong need for research based on LEO communication systems. Currently, research on MIMO LEO satellite communication systems faces several challenges. Firstly, the Doppler shift in LEO satellite channels is much higher than in terrestrial and GEO wireless channels, making reliable channel estimation difficult. Secondly, for the multi-satellite systems and space-ground heterogeneous network systems commonly used in LEO satellites, high-precision networking synchronization for joint MIMO requires multimodal high-latitude delay and angle estimation technologies, which are challenging to apply with existing parameter estimation techniques. Additionally, LEO satellites typically rely on solar panels for power, making their available energy extremely limited. While this book presents an EE optimization method based on lens antenna scenarios, there is still a lack of research on energy optimization in other scenarios. Therefore, many challenges remain in the performance analysis and improvement of MIMO LEO satellite communication systems, warranting further investigation.

MIX
Papier aus verantwortungsvollen Quellen
Paper from responsible sources
FSC® C105338

If you have any concerns about our products,
you can contact us on
ProductSafety@springernature.com

In case Publisher is established outside the EU,
the EU authorized representative is:
Springer Nature Customer Service Center GmbH
Europaplatz 3, 69115 Heidelberg, Germany

Printed by Libri Plureos GmbH
in Hamburg, Germany